LABORATORY MANUAL OF ENTOMOLOGY

LABORATORY MANUAL OF
ENTOMOLOGY

ALKA PRAKASH
M.Sc., Ph.D.
Former Faculty
Department of Zoology
Deshbandhu College (University of Delhi)
Delhi

PUBLISHING FOR ONE WORLD

NEW AGE INTERNATIONAL (P) LIMITED, PUBLISHERS
New Delhi • Bangalore • Chennai • Cochin • Guwahati • Hyderabad
Jalandhar • Kolkata • Lucknow • Mumbai • Ranchi
Visit us at **www.newagepublishers.com**

Published by New Age International (P) Ltd., Publishers
Reprint: 2009

Branches:

- 36, Malikarjuna Temple Street, Opp. ICWA, Basavanagudi, **Bangalore**. © (080) 26677815
- 26, Damodaran Street, T. Nagar, **Chennai**. © (044) 24353401
- Hemsen Complex, Mohd. Shah Road, Paltan Bazar, Near Starline Hotel, **Guwahati**. © (0361) 2543669
- No. 105, 1st Floor, Madhiray Kaveri Tower, 3-2-19, Azam Jahi Road, Nimboliadda, **Hyderabad**. © (040) 24652456
- RDB Chambers (Formerly Lotus Cinema) 106A,1st Floor, S.N. Banerjee Road, **Kolkata**. © (033) 22275247
- 18, Madan Mohan Malviya Marg, **Lucknow**. © (0522) 2209578
- 142C, Victor House, Ground Floor, N.M. Joshi Marg, Lower Parel, **Mumbai**. © (022) 24927869
- 22, Golden House, Daryaganj, **New Delhi**. © (011) 23262370, 23262368

ISBN : 978-81-224-1292-5

Rs. 100.00

C-09-01-3396

Printed in India at Ajit Printers, Delhi.

PUBLISHING FOR ONE WORLD
NEW AGE INTERNATIONAL (P) LIMITED, PUBLISHERS
4835/24, Ansari Road, Daryaganj, New Delhi-110002
Visit us at **www.newagepublishers.com**

PREFACE

The primary objective of this book is to serve as a practical manual for the students of agricultural, horticultural and medical entomology. This book, therefore, provides instructions and working sheets for laboratory work. The instructions for each practical class are presented in such a manner as to enable the students to properly organise their practicals and obtain a better understanding of the fundamentals of applied entomology.

The initial chapters of this book discuss the basic aspects of insect structure, function and development. It is essential to understand these before the student can carry out any practical work. Thus aspects like insect collection and preservation, and pest identification are discussed. While a careful consideration has been given to the requirement of students for facilitating laboratory work, the material is presented in a concise manner. It is suggested that the students supplement other related aspects by any standard textbook on the subject, including **Applied Entomology** by P.G. Fenemore and Alka Prakash.

I would like to acknowledge my gratitude to the various authorities on the subject whose works I have referred to in this book. I am grateful to my students, colleagues and my supportive family, who have all contributed to and encouraged the publication of this book. I sincerely hope that the foundation is laid for further study on the practical aspects of entomology.

ALKA PRAKASH

CONTENTS

1
INTRODUCTION

This book is presented as a working manual for laboratory work in entomology. This manual will complement in your understanding of lectures which are as far as possible given in parallel. Besides providing information and instructions for each practical class, this manual also serves as a laboratory workbook with page space for additional notes and diagrams. It gives you adequate exercises to reconfirm your understanding of the subject. The hands-on activities presented throughout the text provide opportunities for you to get personally involved in entomology and make some of your own observations, rather than just reading from a textbook.

The initial part of the book lists out the dos and dont's for the laboratory work. This is followed by explaining the methods for collecting and preserving the insects. Next two chapters provide the structure-function concepts of both external and internal anatomy of insects. Upon this foundation of study, different types of insect development are discussed briefly. In *Insect classification*, you will find out about many of the insect orders and the differences that set each one apart from the others. This will enable you to recognise the main life stages of a pest species, knowledge of plants attacked and damage symptoms in case of crop pests; and the diseases transmitted in case of animal pests.

Pest identification is an important part of applied entomology. After spotting the causal insect on damaged plants or infested animals, the next step is to conclusively identify the insect to its order. The chapter *Pest Identification* provides two simplified reference keys to insect orders to permit their identification.

Chemical control of insects still forms a major part of integrated pest management programmes in the field. Therefore exposure of students to commonly used terms like LC_{50}, LC_{90} and toxicity of insecticides is essential. Procedures for standard laboratory insecticide experiments form the material for the next chapter. Various types of equipments employed for applying insecticides in the field conditions are discussed briefly in *Insecticide Application Equipment*. A glossary and a selected reference section are provided near the end for use in instructing a class.

2

LABORATORY PRACTICE

A good standard of laboratory practice and behaviour is essential to ensure that equipment is well cared for and for reasons of personal safety. Some materials used in biological laboratories are dangerous if misused (corrosive, flammable, toxic) so make sure that you understand and follow good codes of practice. Any special precautions required will be spelt out prior to commencement of the relevant practical work. Equipments to furnish laboratories is expensive (especially microscopes) so make sure you know how to operate it properly before you use it, otherwise damage can ensue. The following are some important rules of laboratory behaviour. Please observe them.

Laboratory Behaviour

- Do leave bags and other items outside the laboratory or in the area indicated.
- Do handle equipment and materials with care at all times.
- Do put solid wastes into the bins provided.
- Do place used slides in the containers provided.
- Do return your microscope after use.
- Do mop up spillage immediately (take care with possible corrosive materials).
- Do leave your bench area clean and tidy.
- Do put your stool under the bench before you leave.
- Don't eat or smoke in the laboratory.
- Don't leave used slides or specimens on the microscope carriage.
- Don't put solid wastes, solvents or corrosive liquids down the sink.
- Don't pipette by mouth.
- Don't remove specimens, equipment, books or any other material from the laboratory unless you have permission.

Notes on the Handling and Examination of Insect Specimens

Material for examination in practical classes will be provided either at individual work-places or at suitable distribution points around the laboratory. What the material is and where it is located will be explained at the beginning of each laboratory class. *Do not remove material from display areas* or from any other parts of the laboratory.

It is recommended that fresh specimens should be examined as often as possible, because specimens stored in liquid preservatives, or in boxes or cabinets, may lose certain parts. Most insect specimens are fragile and easily damaged, so handle with care at all times. This applies particularly to dried specimens which become very brittle. Always use forceps for smaller specimens rather than fingers.

Most insect specimens have been preserved in alcohol (70% ethanol). All small, soft-bodied insects should be kept immersed in alcohol during examination in the shallow glass dishes (watch-glasses) provided. Larger, robust insects (crickets, beetles, large caterpillars) are best first removed from alcohol and dried before examination.

Always examine specimens first with the unaided eye or with a hand lens and note the actual size. When using the stereomicroscope start with the lowest level of magnification and increase it when the specimen or feature is located. Adjust the direction and intensity of the microscope lamp to provide the best illumination. Select the most appropriate side of the microscope base plate uppermost for contrast with the specimen, i.e., for dark specimens use white side uppermost, for light specimens the reverse.

Notes on Making and Labelling Drawings

Drawings should be simple line drawings to illustrate the features of interest (Fig. 2.1) and not of the high artistic standard. General points are:

1. Make them *large*; one-third to a full page, no thumb-nail sketches.
2. Use a *sharp* pencil.
3. Have a rubber and ruler at hand.
4. Place the specimen in an appropriate position. Remember that it is customary for the head to be at the top for dorsal and ventral views.

STAGE 1: Rough Outline

It is important to get the body proportions correct:

1. Observe the general shape. Mentally note the proportions, e.g., 2 times as long as it is wide.
2. Observe the rough shape and proportions of the main regions in relation to each other, e.g. legs as long as body, head roughly triangular.
3. Lightly outline all the main areas (e.g. head and thorax). Use one or two lines to indicate length and angle of antennae, legs, etc.
4. Check you have space to write the labels around the outside.

STAGE 2: Draw in the Details

1. Use firm lines. Darker for the outlines of the main areas and lighter ones for markings within these areas.
2. Be observant. Don't draw five abdominal segments when there are nine or eyes when there are none.
3. Don't attempt to draw every hair, bump and spine. Consider what the drawing is aiming to illustrate, e.g., if hairiness is an important feature of the specimen, use a series of short light lines to indicate this, otherwise leave them out.
4. Where it is important to indicate fine details:

either (*i*) Draw in a section and use arrows to indicate its extent,

or (*ii*) Draw a separate enlarged illustration alongside. Indicate where it is on the original drawing.

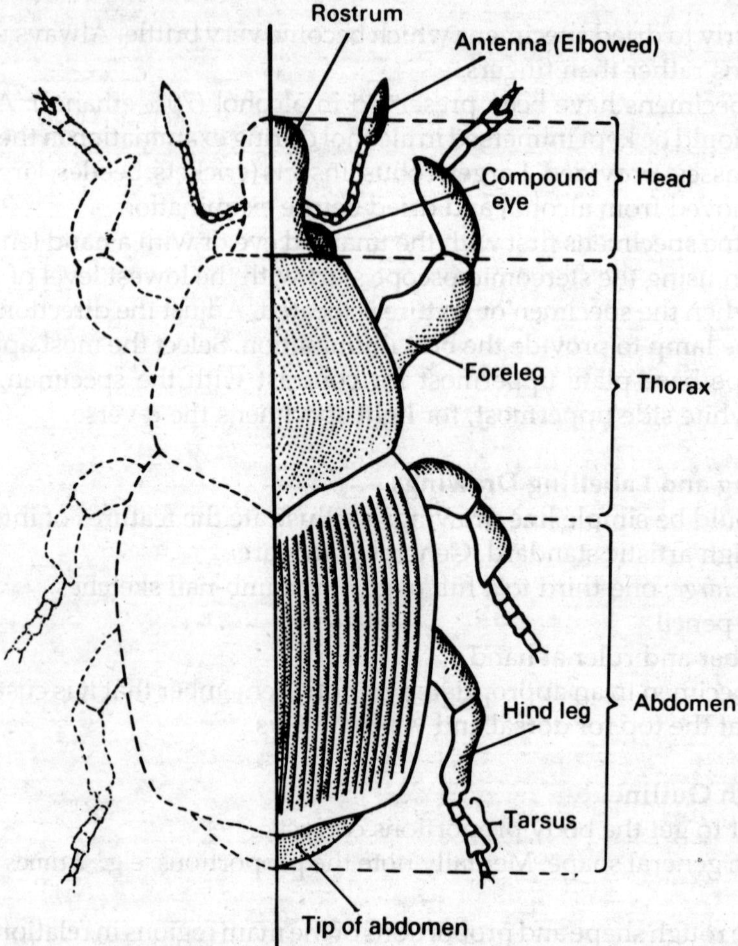

Size : 5mm long. Colour
Stage 1 : Sketch a rough outline
Stage 2 : Draw in the details
Stage 3 : Rule the labelling lines
Stage 4 : Write the labels
Stage 5 : Add the heading and indicate size and colour

Fig. 2.1. A typical adult weevil (example of a simple line drawing of an insect)

STAGE 3: Rule the Labelling Lines

1. Use pencil—easily rubbed out.
2. Rule them from the structure to where you will write the labels.
3. Avoid bends in the lines.
4. Keep them as horizontal as possible.
5. The lines should end on the part being labelled, not just point to it.
6. Avoid crossing labelling lines. Plan before you start.

7. To label several identical parts use 2-3 lines only, all ending at the same label, e.g. for spiracles, segments.
8. To label regions clearly, use brackets and dotted lines.

STAGE 4: Writing the Labels

1. Use pencil.
2. Keep them horizontal as far as possible.
3. Keep them away from the actual drawing.
4. Do not use abbreviations.
5. Don't underline them — it wastes time.
6. Include only labels that are appropriate to the purpose of the drawing. Don't label everything, e.g. you might label 'eye' and 'antenna' in a drawing of "mouth parts in situ on the head".
7. Add notes with labels to clarify points that cannot be seen on the drawing or to which you wish to draw attention, e.g. "wings — two pairs", "antenna (elbowed)".

STAGE 5: Add the Heading and Other Notes

1. These should be put at the *top* or *bottom* of the drawing. Whichever you choose, be consistent.
2. Write them in capital letters or underline or both.
3. Centre them.
4. They should include: words to indicate what the drawing aims to illustrate, the name of the insect, the life stage, the name of the actual species (if known) and the orientation of the drawing; e.g. "Dorsal view of the Head of adult *Periplaneta americana*".
5. Sometimes where there is a series of drawings on one page it is better to have one main heading, e.g. "Life stages of grass grub" and a series of sub-headings on the drawings, e.g. "eggs", "larva", "pupa", "adult".

Other notes: In a suitable place, add if appropriate:

Size: Length is normally measured from head to the tip of the abdomen not including wings, ovipositor, antennae, etc. Give an average measure.

Colour: Indicate colour of live specimens.
Don't forget
—Dried specimens may fade.
—Specimens in alcohol often lose their colour completely.

3

COLLECTION, PRESERVATION AND CULTURE OF INSECTS

NATURE OF INSECT COLLECTION

A good zoological collection should consist of at least four individuals representative of each of the orders of the Insecta. Thus the collection should reflect the great diversity of form which occurs in the Insecta. For the agricultural student, emphasis should be placed on insects of agricultural importance and the collection should reflect the range of pests of the crops of interest. A good place to start collecting insects is a flowering hedgerow, or garden where insects can be found on the flowers, leaves and stems. Another place to look for small insects is in the soil about the roots of plants. Collecting aquatic insects in water—ponds, streams, rivers, lakes—is another side of entomology. Of course, indoor collecting can go on all the year round and besides fleas, bugs and flies, a number of interesting insects can be found.

Collecting Insects

Before proceeding to collect insects, the equipment used for collection should be assembled. The equipment used is simple and inexpensive. The collecting kit includes the following items: killing bottle (Fig. 3.1), pooter [Fig. 3.2 (a) and (b)], hand lens, pen-knife or scalpel, scissors, forceps, insect net [Fig. 3.3 (a) and (b)], beating tray (Fig. 3.4), notebook and pencil, small brushes, stoppered tubes and insect pins.

There are many methods of collecting insects but only important methods and the necessary equipment are discussed here:

1. *Hand picking.* The insects are picked up by hand and placed in a container. This method is used for collecting large insects such as beetles, bugs and grasshoppers. The method is tedious and only a few specimens can be collected easily.

2. *Insect or butterfly net.* A general insect net [Fig. 3.3 (a)] is a light cloth bag hung from a loop that is attached to a handle. The general-purpose net or butterfly net should have a loop of steel wire, 30 cm in diameter and a bag of thin muslin or mosquito netting material. The bag should be tapered towards the bottom and should be atleast twice as long as the diameter of the frame, so that with a twist of the wrist it can be closed over the frame [Fig. 3.3 (b)]. The handle should be about 60 cm long for easy manoevering. The handle may be of wood or metal and is connected by means of a screw arrangement to the metal ring. This type of net is used for catching butterflies, moths, dragonflies,

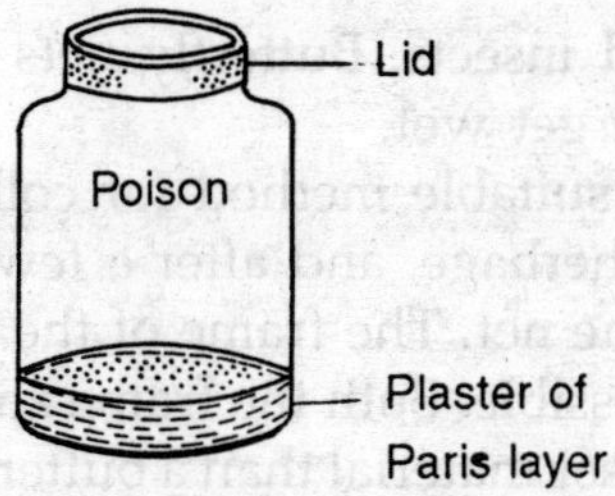

Fig. 3.1. Killing bottle

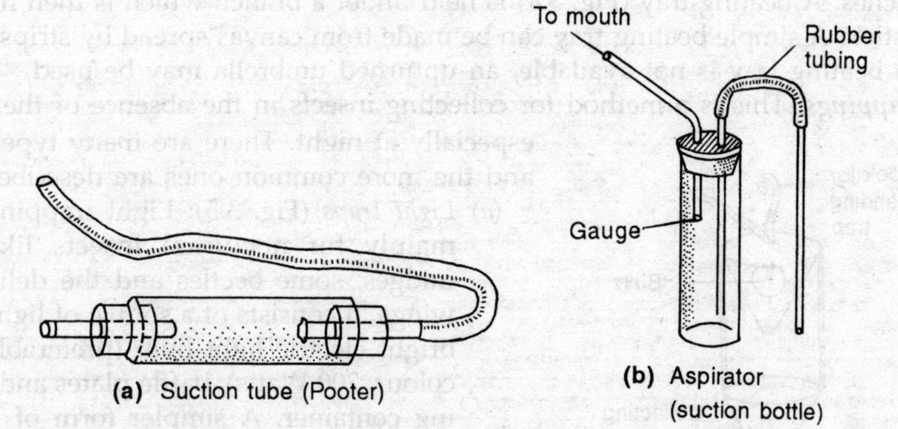

Fig. 3.2. (a and b)

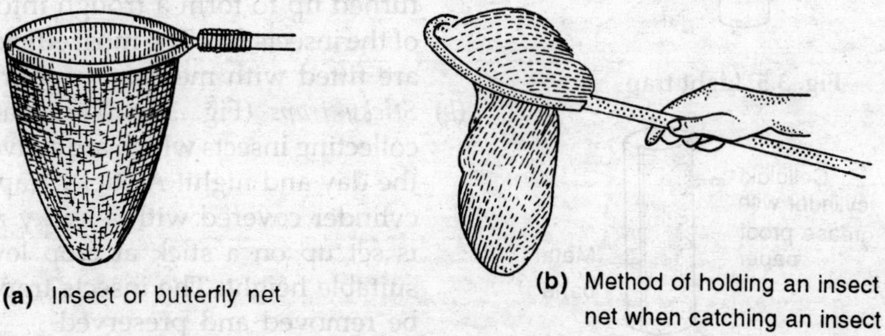

Fig. 3.3. (a and b)

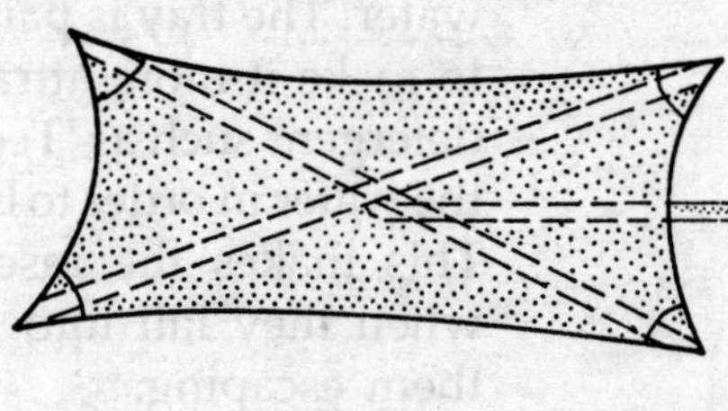

Fig. 3.4. Beating tray

wasps and other large-winged insects. Butterfly nets are best used under sunny day conditions when they will not get wet.

3. *Sweeping.* Sweeping is a suitable method for collecting many insects. The sweep net is used to sweep through herbage, and after a few strokes grasshoppers, bugs and other insects are collected in the net. The frame of the sweep net should be big, to take as big a sample of insects as possible. Both the frame and the bag need to be very strong, hence the bag should be of denser material than a butterfly net, e.g., white calico or heavy muslin.

4. *Beating.* This method is suitable for collecting crawling insects and those which rest on branches. A beating tray (Fig. 3.4) is held under a branch which is then hit sharply with a stick. A simple beating tray can be made from canvas spread by strips of wood. When a beating tray is not available, an upturned umbrella may be used.

5. *Trappings.* This is a method for collecting insects in the absence of the collector, especially at night. There are many types of traps and the more common ones are described here:

(*a*) *Light traps* (Fig. 3.5): Light trapping is used mainly for nocturnal insects like moths, midges, some beetles and the delicate lace-wings. It consists of a source of light, such as bright electric light bulb (preferably of blue colour, 200 Watts), baffle plates and a collecting container. A simpler form of light trap consists of a white sheet hung so that a light shines upon it. The lower edge of the sheet is turned up to form a trough into which some of the insects fall. The most efficient light traps are fitted with mercury vapour lamps.

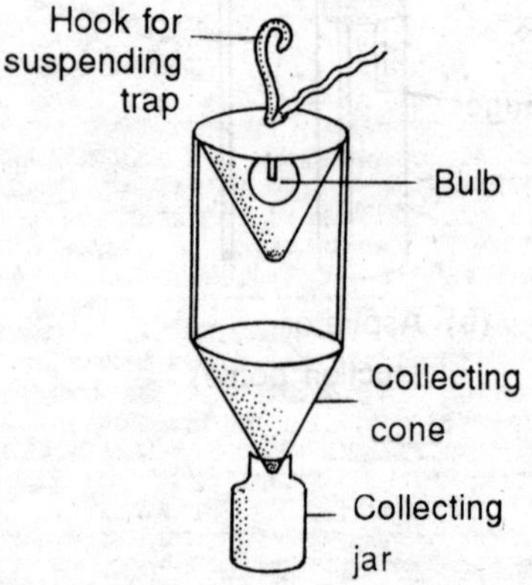

Fig. 3.5. Light trap

(*b*) *Sticky traps* (Fig. 3.6): These are useful for collecting insects which are active during both the day and night. A sticky trap consists of a cylinder covered with a sticky material. This is set up on a stick at crop level or at any suitable height. The insects trapped can then be removed and preserved.

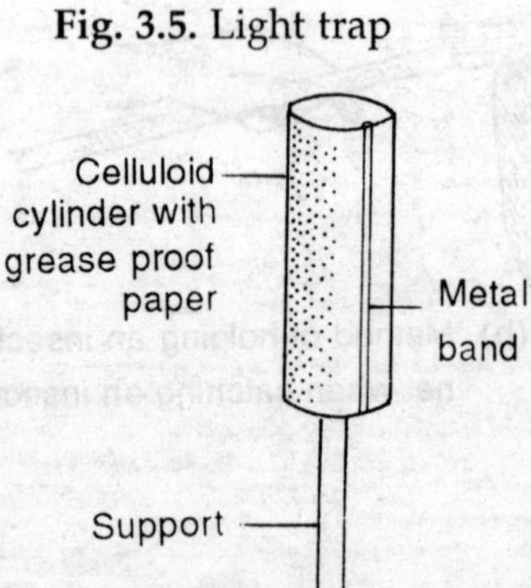

Fig. 3.6. Sticky trap

(*c*) *Water traps* (Fig. 3.7): A water trap consists of shallow tray about 12 × 20 cm containing water. The tray is painted white on the inside to make it more attractive. A few drops of a detergent such as 'Teepol' should be added to the water in order to break the surface tension. This makes the insects sink to the bottom when they fall into the water and prevents them escaping.

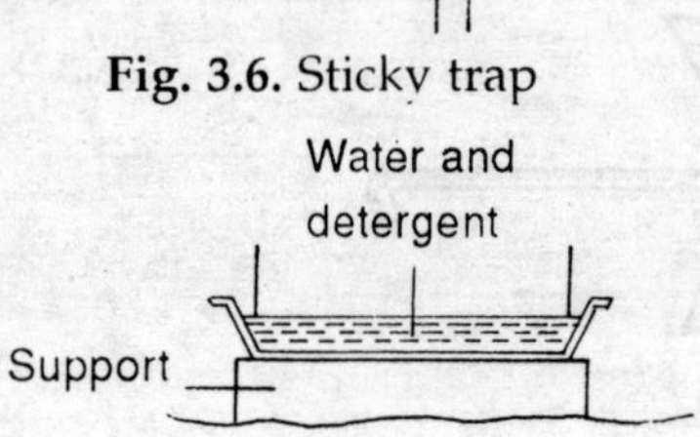

Fig. 3.7. Water trap

(*d*) *Pitfall traps* (Fig. 3.8): Crawling and running

insects like ground beetles may be caught in baited traps sunk into the ground so as to form pitfalls. The pitfall trap consists of a container such as a jar or a rectangular trough which is either open or has a lid with a large central hole. The trap is buried so that the top is at ground level.

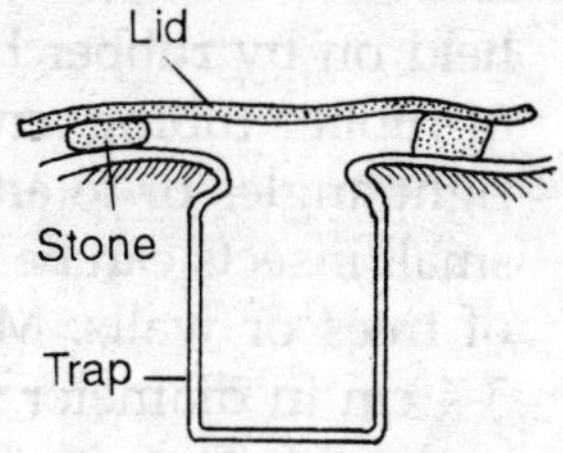

Fig. 3.8: Pit-fall trap

(*e*) *Baits and bait-traps:* Baits are special traps in which an attractive substance is placed which will lure insects into the trap. Sweet and fermenting baits, decaying and over-ripe fruits, honey, syrup, etc. may be used in such traps, and will attract any of the insects that normally visit flowers. A suitable bait for houseflies is decaying meat or fish. Some baited traps are made in the form of wire cages. The general principle is that the insects are attracted to the bait in a shallow container, and then when they rise up after feeding they enter a wire enclosure from which escape is difficult.

6. ***Pond net.*** Aquatic species of insects may be collected with a pond net. A pond net is made of fine nylon netting, usually smaller than a butterfly net but having a long handle. The loop should not be circular, but either square or semi-circular. The net is put into the water at various depths and then pulled quickly to the surface. When the water drains off, the insects left behind in the net are collected.

7. ***Collecting insects from debris.*** A very large number of small insects live in debris of leaf-mould, decaying vegetation, rotten wood, etc., and to capture them a sifter is used. Any container with a wire-mesh bottom will serve as a sifter. The ground litter is put in the sifter and shaken gently over a piece of white cloth. As the insects fall onto the cloth they may be easily captured with an aspirator or forceps. A modification of this is a separator, usually called a Berlese funnel. This consists of a funnel over which a sieve containing the litter may be placed. Immediately over the sieve an electric lamp is fitted. The funnel leads into a receiver containing liquid preservative. As the debris dries the insects gradually move downwards to the conditions that they prefer and fall down the funnel into the preservative. The heat must be applied gradually, so that the insects are not harmed.

Transferring to Containers—Getting the Insects out of the net

Aquatic insects are generally fairly helpless when the water is drained away, and then can be tipped into a container or carefully picked up with forceps. Aerial insects are, however, more active and difficult to pick them up when alive. Use may be made of the following:

(*a*) *Tubes*, glass or transparent plastic, with corks to fit.

(*b*) *Collecting bottles.* This is a variation of the tube in which both ends are open and one end leads to a bottle.

(*c*) *Aspirators* or *suction bottles* or *pooters*: These are used to collect small insects from a net or occasionally directly from foliage, walls or from the ground. A suction bottle [see Fig. 3.2 (a)] is made by taking a small bottle and fitting the cork with two small tubes instead of one. One tube has a small cork at its outer end and its inner end open; the other tube is closed at its inner end with a piece of muslin, held on by rubber band, while the outer end of this tube is fitted with a length of rubber tubing, ending sometimes in a mouthpiece. The two tubes are bent at right angles or to any other angle of convenience. By sucking at the rubber tube, small insects can be drawn into the bottle from the net or directly from the trunks of trees or walls. Many collectors prefer a straight tube [see Fig. 3.2 (b)] about 3-4 cm in diameter (pooter) in place of the bottle. This has two smaller tubes at each end. The chief advantage of the suction-tube over the suction bottle is that both ends can be opened for cleaning.

Killing Insects

The method of killing insects may have an important effect on their permanent preservation and this should be kept in mind when deciding which method to use.

Killing bottles: Any wide-mouthed glass jar such as an empty jam jar, can be used to make a killing bottle (see Fig. 3.1). The jar should have an air-tight lid. The bottom should be lined with a layer of plaster of Paris into which potassium cyanide, ethyl acetate, chloroform, ether, tetrachloroethane or any suitable poison is added. Cyanide killing bottles are very dangerous to keep in the house, so are best avoided. Ethyl acetate has been found safe and efficient to use. A temporary killing bottle can be made by putting a piece of cotton wool soaked with ethyl acetate in a wide-mouthed bottle. If any liquid remains free on the surface it will wet the specimens and spoil them. Insects should not be left for more than 20 minutes in the killing bottles, otherwise they become discoloured. Dispose of the bottles that no longer kill by burning or burying.

Other Methods of Killing Insects

Pinching: The bigger insects, particularly butterflies and moths can be killed by pinching the thorax between thumb and finger in one quick, decisive movement. Its advantage is that it kills the insect instantaneously, without giving it a chance to struggle and break its wings or knock off the scales. The correct amount of pressure to apply without damaging the insect is learned only with practice.

Injecting: Butterflies and moths can also be killed by injecting one of the killing fluids or oxalic acid into the thorax with a hypodermic needle. This kills them instantly.

Hot water (not boiling) can be used at home or in the laboratory. Larvae are dropped alive into a small quantity of hot water, which is then allowed to cool before they are transferred to 80 per cent spirit.

It is generally recommended to mount and set insects within 24 hours of killing them, otherwise the joints will become stiff.

Preserving Insects

Insects can be preserved either for temporary storage or for permanent collections in boxes and cabinets.

Temporary Storage

In this method insects are enclosed neatly between two layers of cotton wool or cellulose wadding, with the relevant data written on a card, and wrapped in an envelope of old newspaper. This is known as *layering* and is the most useful way of storing dead insects temporarily (up to a few years). Some drops of paradichlorobenzene or carbon tetrachloride, or flakes of naphthalene, should be added to the cellucotton to protect the insects from attack by beetles. Many such packets can be stored in a wooden box in such a way that neither the specimens shift about nor are pressed down to distort them.

Papering is another way of temporarily protecting soft-bodied insects that have large wings, or long legs like butterflies, moths and craneflies. Papering consists of enclosing the specimen in a triangular envelope. To make an envelope, the paper is cut into rectangles with their sides in the proportions of 3 : 5. The method of folding is shown in Fig. 3.9 (a) and (b), and it produces an envelope in the shape of an isoceles triangle. Only one specimen should be put in each paper.

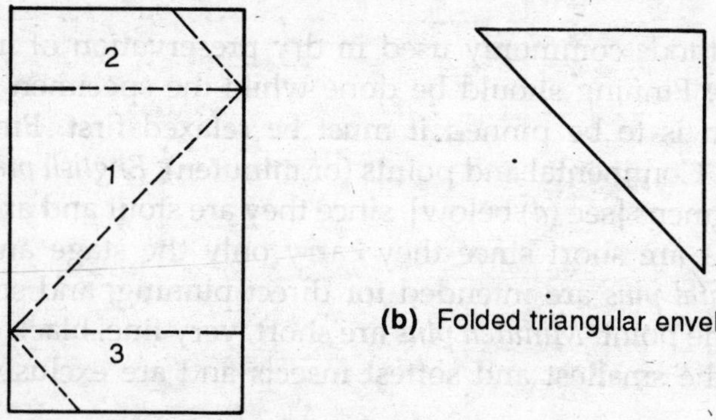

(a) Method of folding a rectangular piece of paper to form a triangular envelope

(b) Folded triangular envelope

Fig. 3.9

Permanent Preservation

Methods of permanent preservation fall into three categories: preserving the specimens dry; keeping them in a liquid; and mounting on a microscope slide.

Dry preservation. Pinning and drying is the most common and a useful method to render most hard bodied insects permanently available for study or display. After the insects have been killed in a killing bottle, they are spread on a pinning stage and a rustless entomological pin is pushed through the thorax. The pinned specimen is kept in a drying chamber for a few days so that it is slowly dried and thus becomes well-preserved. However, in collecting specimens, they often become hard and brittle after a day or two. Before pinning such insects, a technique of re-softening or *relaxing* of the external skeleton is essential.

Relaxing can be done by keeping the specimen in a humid atmosphere until the integument absorbs enough moisture to become soft again. To make a relaxing jar, wide-mouthed jars or tins are taken with tightly fitting lids and a thick layer of sand or any absorbent material placed in the bottom. The absorbent layer is moistened with plain water and a few drops of phenol or ethyl acetate are added to it to prevent fungus growth. It is better to have an upper layer of cork or blotting paper so that the moisture is not directly in contact with the insects. An exposure of one to three days will relax most specimens.

When insects of different orders are collected in the same jar, *cleaning* sometimes becomes necessary. It is best done on relaxed specimens. Solid matter, scales, pollen dust can be picked off with a small camel-hair brush. More fragile insects are immersed in cleaning fluid for some time and the dirt dislodged after wetting with a soft brush. Greasy specimens, which have considerable fatty tissue, should be degreased before mounting. Degreasing the specimens requires them to have a bath in an organic solvent like ethyl acetate, benzene or ether till the fluid ceases to become yellow from the dissolved oils. After cleaning, those specimens that have been treated in water or weak alcohol should be returned to the relaxing jar. Specimens cleaned in the volatile fluids can be simply dried in air.

The pinning methods commonly used in dry preservation of insects are as follows:

(*a*) ***Direct pinning:*** Pinning should be done while the specimen is still fresh, but if a hardened specimen is to be pinned it must be relaxed first. Pins fall into 3 general categories, English, Continental and points (or minuten). *English pins* are preferably used in staging the specimens [see (*d*) below], since they are stout and are handled nicely with forceps. These pins are short since they carry only the stage and the labels, not the specimen. *Continental pins* are intended for direct pinning, and so they are longer and thin and sharp at the point. *Minuten pins* are short, very fine, black or stainless steel pins, used for pinning the smallest and softest insects and are exclusively used for staging (see below).

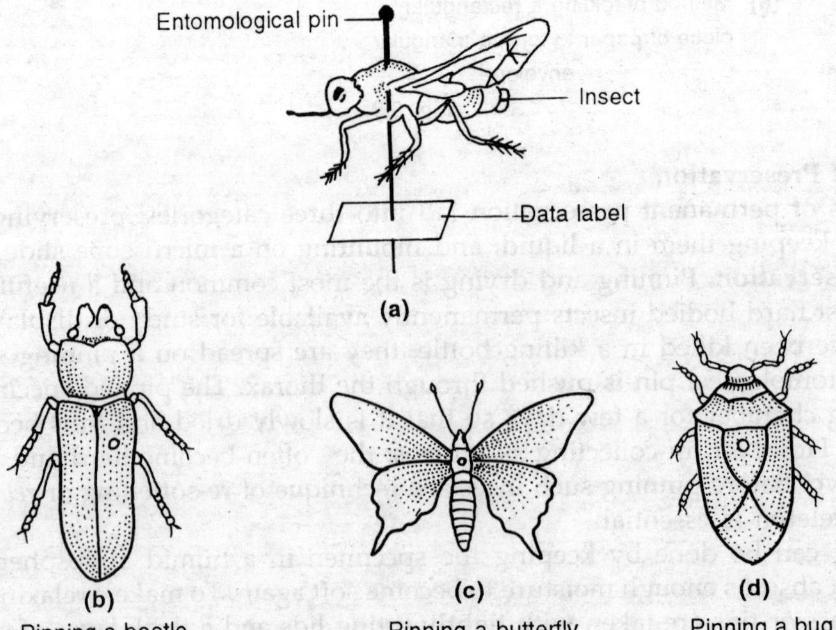

Fig. 3.10.

In direct pinning the insect is mounted by inserting a pin symmetrically positioned through the thorax and all the labels, in such a way that about a quarter of the pin protrudes above the insect but still has enough length to hold firmly into the bottom of the box [Fig. 3.10 (a)]. The pin must be sharp as well as long, to go through everything cleanly. Large heteroptera (bugs) are pinned through the mesoscutellum while large coleoptera (beetles) should be pinned through the right elytron [Fig. 3.10 (b), (c) and (d)].

(*b*) ***Carding:*** The insect is placed on a white rectangular card, measuring 5 × 8 mm or 5 × 12 mm, depending upon the size of the specimen (Fig. 3.11), which is stuck with a drop of gum (such as clear cellulose acetate cement). The antennae and appendages are then arranged well and stuck to the card. An entomological pin is passed through one end of the card and the data labels are attached underneath the card. This method is suitable for small Coleoptera, Heteroptera and ants.

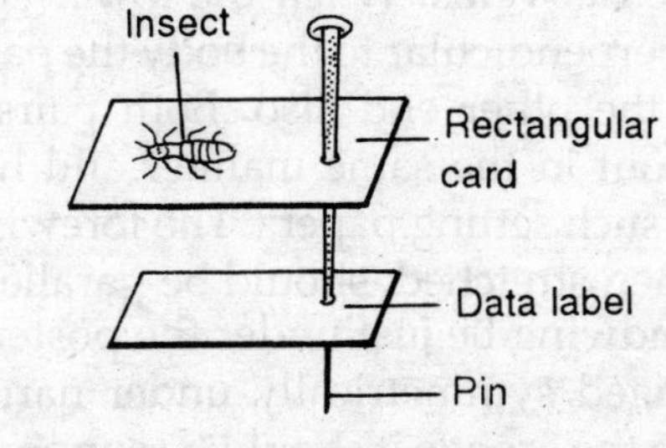

Fig. 3.11: Carding with rectangular card

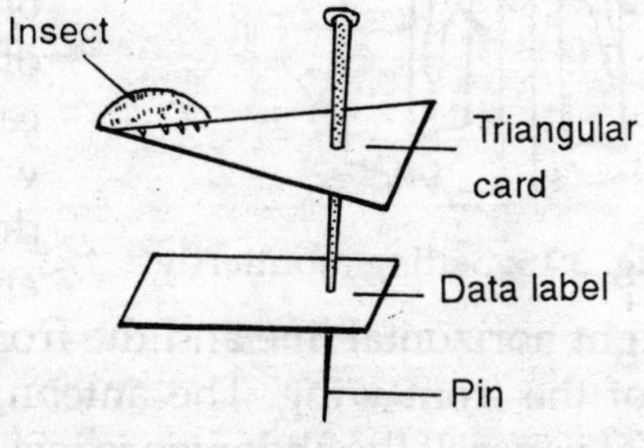

Fig. 3.12: Carding with triangular card

(*c*) ***Triangle carding or pointing:*** This is the best way of mounting the smallest dry specimens. A triangular card point is cut out of Bristol board, and a small drop of gum is placed at the apex of the triangle (Fig. 3.12). The tip of the card point should be bent down at a slight angle so that when the insect is in an upright position the bent tip of the point fits against the side of the insect. This bent tip is attached to the right underside of the thorax.

(*d*) ***Micro-pinning or staging:*** This method is used for minute insects such as small Heteroptera, Hymenoptera and Diptera. It is a type of double mount in which a minute stainless steel pin is pushed through the thorax into a white polyporous support or stage (Fig. 3.13). A larger pin is then pushed through one end of the polyporous.

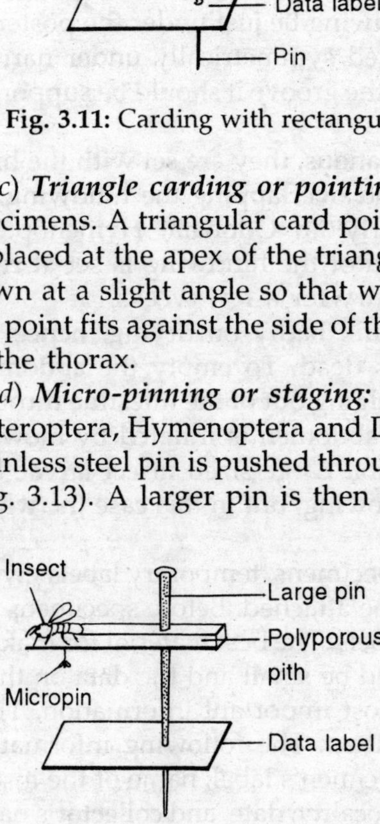

Fig. 3.13: Micro-pinning or staging

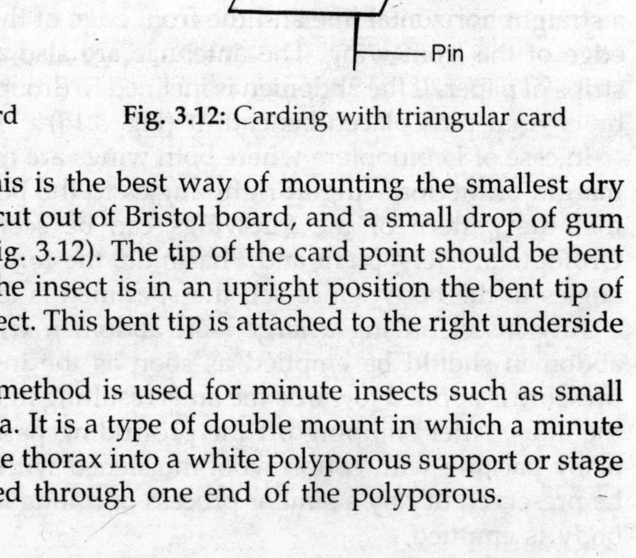

Fig. 3.14: Setting board

(e) ***Setting or spreading:*** Setting an insect means to spread the wings and legs in a horizontal position on a standard setting board. Setting boards can be bought ready-made, or easily made locally. The setting boards (Fig. 3.14) have a groove in the middle to take the body of the insect and this groove has underneath it a sheet of cork. A simple method of making a setting board is to take a thick sheet of cork and cut a groove in it for the body of the insect. The specimen is thoroughly relaxed, and pinned through the body on a long pin of appropriate thickness. The insect is so placed that the thorax and abdomen rest in the groove of the board. A strip of setting-paper is then pinned at the front end, and the forewing is drawn forward to the desired position by inserting a fine pin behind one of the strong costa veins. When the lower edge of the forewing is perpendicular to the body the paper can be pinned at the other end also. Both pairs of wings are spread out in the same manner and held down by means of such setting papers. The forewings and hindwings when stretched, should be parallel in a straight horizontal line and the front edge of the hindwing be just under the posterior edge of the front wing. The antennae are also arranged symmetrically under narrow strips of paper. If the abdomen is inclined to droop into the groove it should be supported by crossed pins placed beneath it (Fig. 3.15).

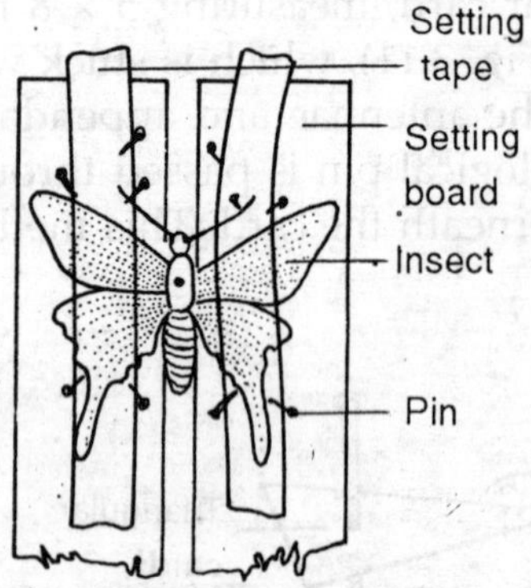

Fig. 3.15: Setting a butterfly

In case of Lepidoptera where both wings are membranous, they are set with the hind margin of the forewing at right angles to the body and overlapping the hindwing, so that the pattern of the forewings can be seen clearly. In Odonata, Hymenoptera, Orthoptera, Dictyoptera and Phasmida, the foremargin of the hindwing is set at right angles to the body. Once set, the specimen is left to dry for a few weeks.

Any insect having a large, soft abdomen will shrink badly on drying, hence the abdomen should be emptied as soon as the insect is dead. To empty the abdomen, pressure is applied towards the anus resulting in the bulging out of the intestine through the anus. After snipping off the protruding part, the abdomen is inflated by blowing or by packing with cotton wool moistened with phenol. Large-sized insect larvae can be preserved dry by a similar process of rolling and blowing, but in this case the whole body is emptied.

(f) ***Labelling:*** During preparation and mounting of specimens, temporary labels giving information about date and place of collection should be attached. Before specimens are put away in storage they should be given permanent labels. The best material for making permanent labels is thin white card. These labels should be small and the data on them must be restricted to the most important information. They can be printed or hand-written. The following information should be given on each specimen's label: name of the insect, crop attacked or host plant, locality, date, and collector's name (Fig. 3.16). The writing ink should be permanent and should not be spoiled if the labels are put in liquid preservative.

P. demoleus
Citrus sp.
New Delhi
24.5.1996
A. Prakash

Fig. 3.16: Making a label.

Permanent collections of pinned insects are kept either in store boxes or in cabinets of drawers. To make such insect boxes pestproof, a small amount of naphthalene, camphor or paradichlorobenzene may be included, either in a cloth bag or a small box firmly pinned in the corner. Naphthalene bolls may be pinned in the box by attaching the boll to an ordinary pin. To do this, head of the pin is heated and then forced into the boll.

Liquid Preservatives

Larvae and nymphs of most insects, and soft-bodied adults are best stored permanently in a liquid. Liquid preservation is also suitable for keeping duplicates of the large numbers of insects obtained from light traps. The most useful fluid preservative is spirit—75-80 per cent ethyl alcohol. Formalin is generally not recommended as a preservative for insects because it hardens the tissues. A few drops of glycerine are added to the alcohol to prevent stiffening. The tubes should be well stoppered and examined periodically to be sure that the specimens do not become dry. The label should be written with a pencil or Indian ink and then dropped into the tube. The tubes are stored in jars having straight side walls and close-fitting lids. The tubes should be put only round the outside of the jar with their labels visible and the middle of the jar packed with cotton wool so that all labels can be read without removing the tubes.

Grades of alcohol can be made from rectified spirit as follows:

Spirit	*Quantity of distilled water*	*Grade of alcohol*
10 vols.	1.5 vols.	80%
10 vols.	3.0 vols.	70%
10 vols.	8.5 vols.	50%
10 vols.	20.5 vols.	30%

Most insect larvae are preserved in 75-80 per cent ethyl alcohol as mentioned above, but several other preservatives may be used. The following are the combinations most often used:

1.	95% ethyl alcohol	100 parts
	Distilled water	100 parts
	Formaldehyde	13 parts
	Glacial acetic acid	5 parts
2	95% ethyl alcohol	55 parts
	Distilled water	35 parts
	Formaldehyde	5 parts
	Glacial acetic acid	5 parts

3	95% ethyl alcohol	15 parts
	Formaldehyde	6 parts
	Glacial acetic acid	2 parts
	Distilled water	30 parts

Mounting—Preparation of Permanent Slides

Small insects such as scale insects, aphids, lice, thrips and others or parts of insects, can be satisfactorily studied only after they are mounted on a microscope slide. The following procedure should be adopted to make good slides:

(*a*) ***Maceration:*** For preparation of sclerotised parts of insects, boil in 5 per cent or 10 per cent potassium hydroxide solution over a low flame. For delicate specimens, such as silver fish or mosquitoes, soaking them in cold caustic potash is sufficient instead of boiling. The material gradually becomes brownish, and sometimes distended. When sufficient softened, remove and wash the material in plenty of water. Then transfer the specimen to water containing a little glacial acetic acid.

(*b*) ***Dehydration:*** This is done by transferring the specimen from acid water into grades of alcohol, starting from 30 per cent, through 50 per cent, 70 per cent, 90 per cent to absolute alcohol, for 10-20 minutes each depending on the size of the material. During each transfer, the specimen must be pressed with a seeker to ensure that the alcohol penetrates all the parts of the insect.

(*c*) ***Clearing:*** The dehydrated specimen is now transferred to a clearing fluid-like clove oil, cedar wood oil or xylene. Generally, xylene is used for clearing and before the specimen is put into xylene, alcohol is blotted up with tissue paper. After 15-20 minutes, the material becomes somewhat transparent and is ready for mounting. The specimen must not remain too long in xylene, otherwise it becomes brittle.

(*d*) ***Mounting:*** Most commonly used mountants are 'Depex', Canada balsam and euparal. To mount the specimen, clean the slide thoroughly with absolute alcohol and spread a small quantity of mountant on the middle. Then place the specimen on the mountant, arrange it and allow to dry for about 10 minutes. Then add a further drop of mountant to the material. Lower the cover-slip down gently, using a pin to support one edge. Do not drop the coverslip flat with the fingers because this will trap air bubbles.

When making thick preparations put chips of glass, parallel to the sides of the slide and stick in place with mountant. Special mounting rings can also be used, but it should be remembered that the ring and the coverslip must be of the same size.

(*e*) ***Finishing:*** To make the slides more permanent, they should be 'ringed' with a suitable varnish after drying. It is easier to ring slides with circular coverslips than those with square or rectangular ones.

The slide should be labelled fully as soon as it is made, with a white paper label stuck across one end. Labelled slides should be stored in suitable slide boxes.

Culture of Insects

Culturing of insects involves all the three activities, of *keeping, rearing* and *breeding insects*. Most insects can be kept alive for some time after capture and the larvae or pupae can usually be induced to complete their development. However, breeding means maintaining them through a complete life-cycle, mating, oviposition, successive larval moults, pupation and emergence of adults. Insects that can readily be bred continuously in the laboratory are those whose immature stages are all passed in a uniform and stable medium.

Keeping adult insects alive. The insect is put in a glass jar in which some sand or soil is placed at the bottom, and covered on top with muslin or netting. Insects need enough moisture, protection from extremes of temperature and suitable food. Caterpillars, bugs and other plant feeders should have a supply of the plant on which they were caught. Raisins are a useful supply of food and moisture for many insects. The raisins should be chopped or pulled apart, dropped into the cage and renewed if they show signs of mould.

Rearing insects. Insects may be captured in an immature stage—nymph, larva or pupa, and reared to adults. To rear insects successfully, the natural conditions under which the immature insects were found should be simulated as closely as possible in the rearing cages. Rearing is easy to carry out with most insects which infest stored products or those which live in dung, decaying vegetable matter or rotting fruit. Insects that feed on living plants may be caged over potted plants or fed frequently with fresh material from their host plant. Carnivorous larvae, such as those of the blow-flies and mosquitoes, can be reared on an entirely artificial medium consisting of dog biscuits, agar, baker's yeast and salt. Suitable sites, like sand or a twig need to be provided in the rearing cage for pupation. Moisture and temperature regulation are crucial during pupation and most pupae need moisture when they are getting ready to emerge. Butterflies, moths, beetles and many other insects can be obtained by collecting pupae and maintaining them till adult emergence. Rearing cages should always be protected from ants and other predators.

Breeding insects. The place where insects are bred and maintained under laboratory conditions is known as an insectary. It may consist of one small room, in which both larvae and adults are reared, or several rooms. As an example, we shall consider breeding of mosquitoes in the laboratory.

Eggs of mosquitoes are obtained by isolating gravid females in a mosquito cage, in petridishes lined with a thick layer of damp cotton wool which is covered with filter paper. The eggs are allowed to hatch in the container in which they were laid. A day after hatching, larvae are transferred to white polyethene rearing pans filled with dechlorinated water. About 200 larvae are put in each pan and food in the form of dry powder is given. Food consists of mixture of brewer's yeast and dog biscuits, finely powdered. The larvae should be fed with small quantities of food at a time to avoid fermentation and development of fungus. Total larval period is 8-19 days at 28°C. After the larvae complete their development they pupate in the rearing pan. The pupae are separated from the larvae and are placed in the breeding cages. The majority of the adults will emerge in about two days. Adults feed on soaked raisins or sugar solution. Female mosquitoes require a blood meal at intervals of 2-3 days which is provided by rabbits, guinea pigs or rats. To maintain the colony the females have to be fertilized by natural or artificial mating. In most of the cultured or artificially bred species of mosquitoes, natural mating in captivity occurs. In this way, a continuous breeding colony of mosquitoes can be maintained in the laboratory.

4

EXTERNAL ANATOMY OF INSECTS

Two insects will be examined for major external features; an adult of black field cricket (order Orthoptera) and a larva (caterpillar) of a moth (order Lepidoptera). The specimens are provided in alcohol (70% ethanol). They should be removed and dried gently with paper tissue before examination. Replace the specimens in alcohol when finished with.

WHOLE ADULT INSECT (Fig. 4.1)

1. Body is divided into three regions: head, thorax (bearing legs and wings) and abdomen.
2. Segmented nature of the body. The segments show clearly in the abdomen.
3. Hardened nature of the exoskeleton, especially in the head, thorax and legs. Movement is permitted by soft membranous areas between the hardened parts.

1. THE HEAD

1. Structure: The head is entirely encased in rigid exoskeleton (head capsule) with no clear signs of segmentation. The head capsule, however, bears numerous sutures or lines which correspond to the internal strengthening ridges, which occupy relatively constant positions and delimit to some extent the morphological areas into which the wall of the head can be divided (Fig 4.2 a and b). Besides, the head of immature insects is crossed by a so-called *ecdysial cleavage line* (inverted Y-shaped) on the antero-dorsal surface.

2. Orientation: Based on the inclination of the long axis and the position of mouth parts, the heads of insects are orientated into three types.

(a) *Prognathus:* The long axis of the head is horizontal, or slightly inclined ventrally and this is in line with the long axis of the insect's body. The mouth parts are pointed forwards, e.g., stick insects (Fig. 4.3).

(b) *Hypognathus:* The long axis of the head is at right angles (i.e., vertical) to the long axis of the body. The mouth parts are directed downwards ventrally, e.g., crickets, grasshoppers (Fig. 4.4).

(c) *Opisthognathus:* Although the long axis of the head is horizontal, the mouth parts are directed backwards so that they arise between the anterior legs, e.g., plant-sucking bugs (Fig. 4.5).

3. The compound eyes: These are large and well developed with many small

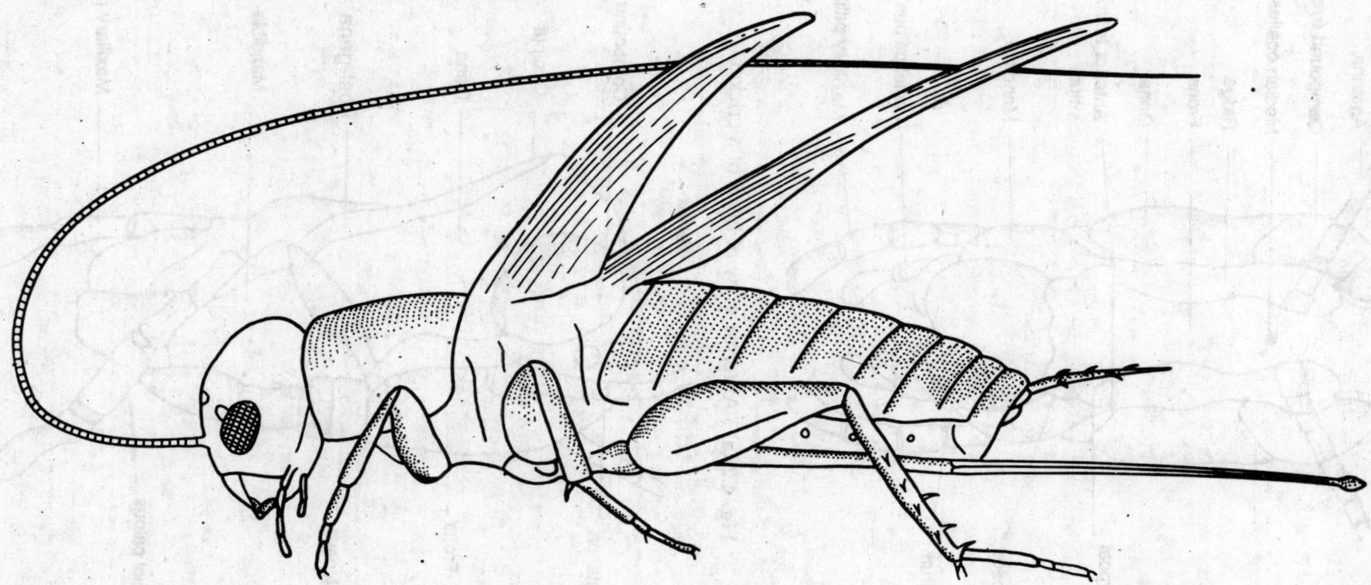

Label the following on the drawing:
Head, Thorax, Abdomen, Prothorax, Compound eye, Ocellus, Antenna, Mouth parts, Mandibles, Palp, Forewing, Hindwing, Body segments, Spiracles, Cercus, Ovipositor

Length of insect (excluding ovipositor) to nearest mm. ____________

Fig. 4.1

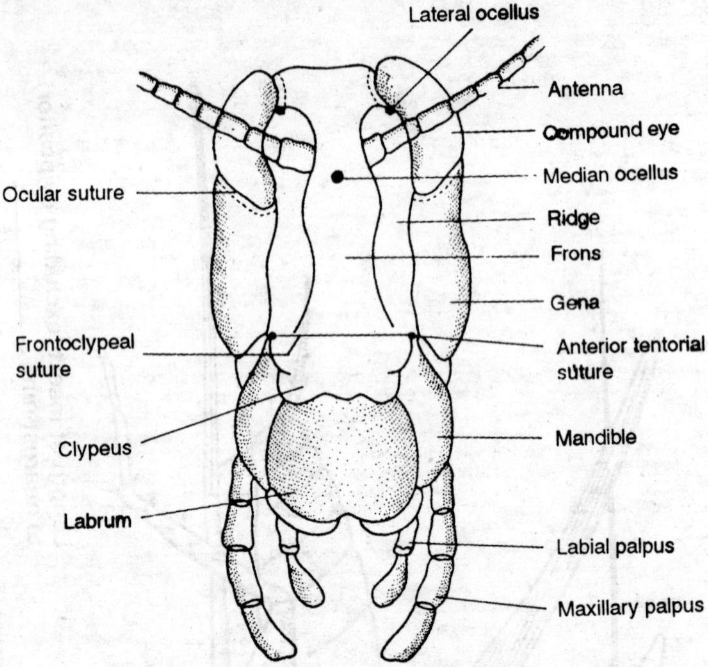

Fig. 4.2 (a) : Anterior view of head of a cricket.

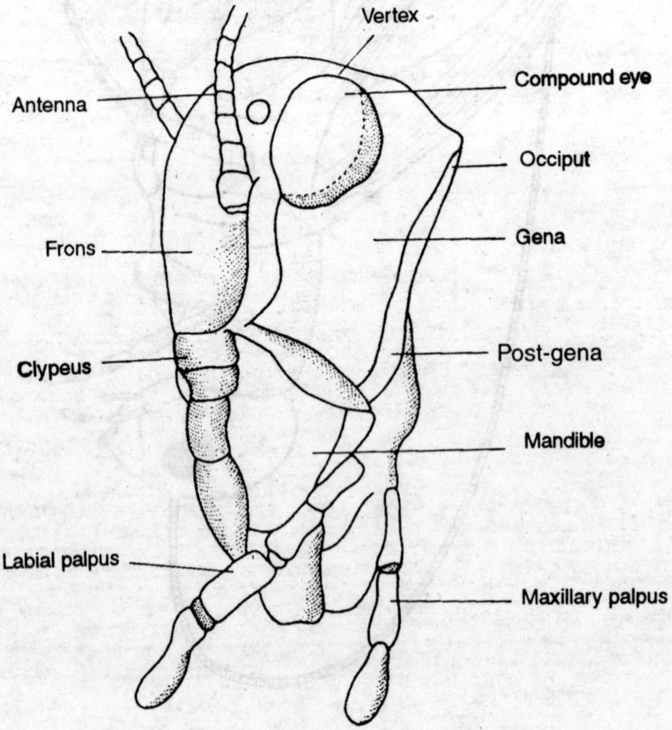

Fig. 4.2 (b) : Lateral view of head of a cricket.

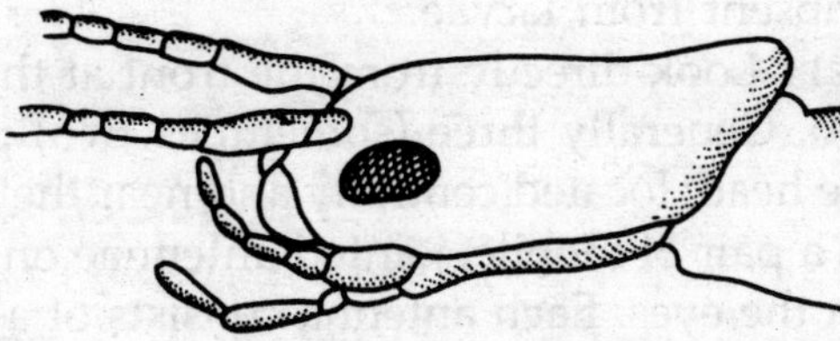

Fig. 4.3 : Prognathus head-stick insect.

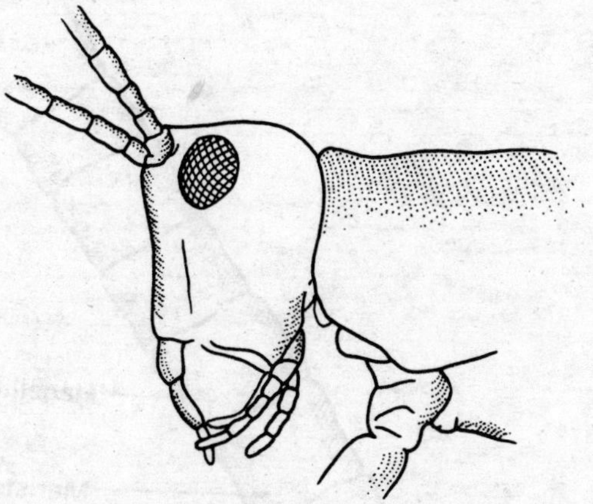

Fig. 4.4 : Hypognathus head-grasshopper.

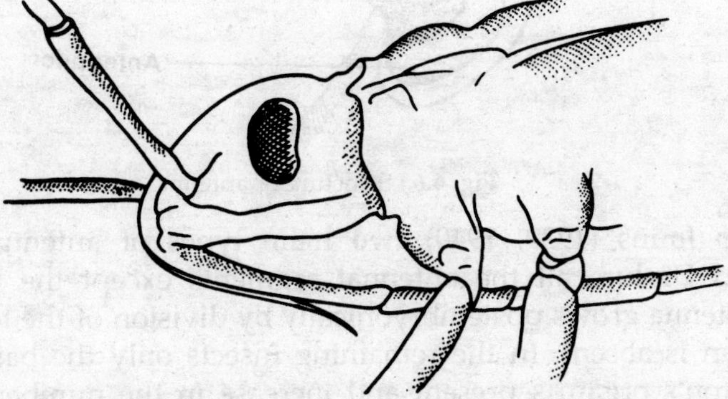

Fig. 4.5 : Opisthognathus head-cotton bug.

individual facets (use highest magnification of a stereomicroscope). Compound eyes are present in most insects but absent from larvae.

4. The Ocelli (simple eyes) : Look directly from the front at the "face" of the insect using moderate magnification. Generally three (sometimes two) pale patches are seen set into the dark surface of the head; located centrally and near the base of each antenna.

5. The Antennae : There is a pair of mobile, jointed antennae on the head of an insect arising in front of or between the eyes. Each antenna consists of a basal segment called the **scape,** followed by the **pedicel** and the remaining part is called **flagellum**. The scape is inserted into the membranous region of. the head wall, the *antennifer* (Fig. 4.6). Antennae are sensory in function (touch and smell). Besides, they can detect humidity changes, vibrations, wind velocity and direction. Antennae are absent in Protura, and they are practically absent in most larvae of the higher Hymenoptera.

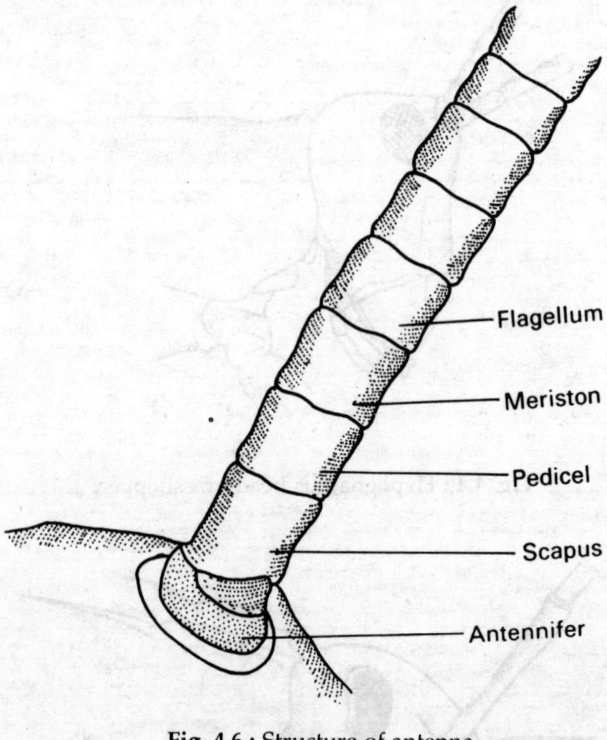

Fig. 4.6 : Structure of antenna.

According to Imms (1939, 1940) **two** main types of antennae are there. In the Collembola and Diplura all the antennal segments except the last contain intrinsic muscles, the antenna grows postembryonically by division of the terminal segment and Johnston's organ is absent. In the remaining insects only the basal segment contains muscles, Johnston's organ is present and increase in the number of segments occurs through division of the third segment or sometimes of the more distal segments. There are numerous variations in the antennae of different orders and families of insects, which can be used to distinguish one from the other. Examine and draw different types of insect antennae (Fig. 4.7 and Table 4.1).

TABLE 4.1: Types of Antennae

Type	Description	Found on
(i) Filiform	thread-like	field cricket (Fig. 4.7a)
(ii) Setaceous	bristle-like	cockroach (Fig. 4.7b)
(iii) Capitate	head-shaped	nitidulid beetle (Fig. 4.7c)
(iv) Clavate	tapering club-like	butterfly (Fig. 4.7d)
(v) Lamellate	flat plate-like	June beetle (Fig. 4.7e)
(vi) Moniliform	string of beads	termites (Fig. 4.7f)
(vii) Serrate	saw toothed	click beetle (Fig. 4.7g)
(viii) Geniculate	elbowed	chalcid (Fig. 4.7h)
(ix) Pectinate	comb-like	moth (Fig 4.7i)
(x) Plumose	with plumes	mosquito (Fig. 4.7j)
(xi) Aristate	with an arista	housefly (Fig 4.7k)
(xii) Flabellate	pile of thick plates	cedar beetle (Fig. 4.7l)

6. The mouth parts : The group of structures on the lower part of the head are known as mouth parts, which are adapted to their diet. The mouth parts of insects are formed from modified appendages of segments making the head. They surround the mouth opening and in most cases are quite exposed and visible. However, in a few insect groups, e.g., Collembola the mouth parts are enclosed by the lower part of the head.

Usually insects have mouth parts adapted for one of three modes of feeding:

(a) Biting and chewing

(b) Piercing and sucking

(c) Rasping.

The structures associated with these modes of feeding are quite different in each case. In addition, many other types of mouth parts occur, specialised for other feeding purposes. A range of these will be examined to give some idea of the variety of structure that can occur in other insects.

(a) **Biting and chewing mouth parts (Orthoptera) :** The basic pattern of biting and chewing mouth parts is found in the black field cricket (very similar also in grasshoppers, locusts and cockroaches). Examine the intact head of the cricket from the side and identify the following parts with the aid of the Figs. 4.8 and 4.9. From front to rear :

(i) Labrum (single): This is flap-like "front lip" and is attached to the lower part of the clypeus. It forms a protective cover over the front of the mandibles.

(ii) Mandibles (paired): These are situated immediately behind the labrum, meeting in the mid-line. They are heavy biting jaws, serrated and black towards the inner edge, hinged to the lower edge of the subgena.

(iii) Maxillae (paired): These are "supplementary jaws", situated behind and slightly out to the side of each mandible. Each maxilla bears a long maxillary palp (sensory in function) and a cardo, stipes, inner lacinea and outer galea.

(iv) Labium (single): This is the "hind lip" or lower lip. It is formed by the fusion of two similar parts and consists of submentum, mentum, prementum, glossa and paraglossa. Labium bears two palps which are slightly shorter than maxillary palp.

(v) Hypopharynx: There is an additional fleshy tongue-like part to the mouth parts that

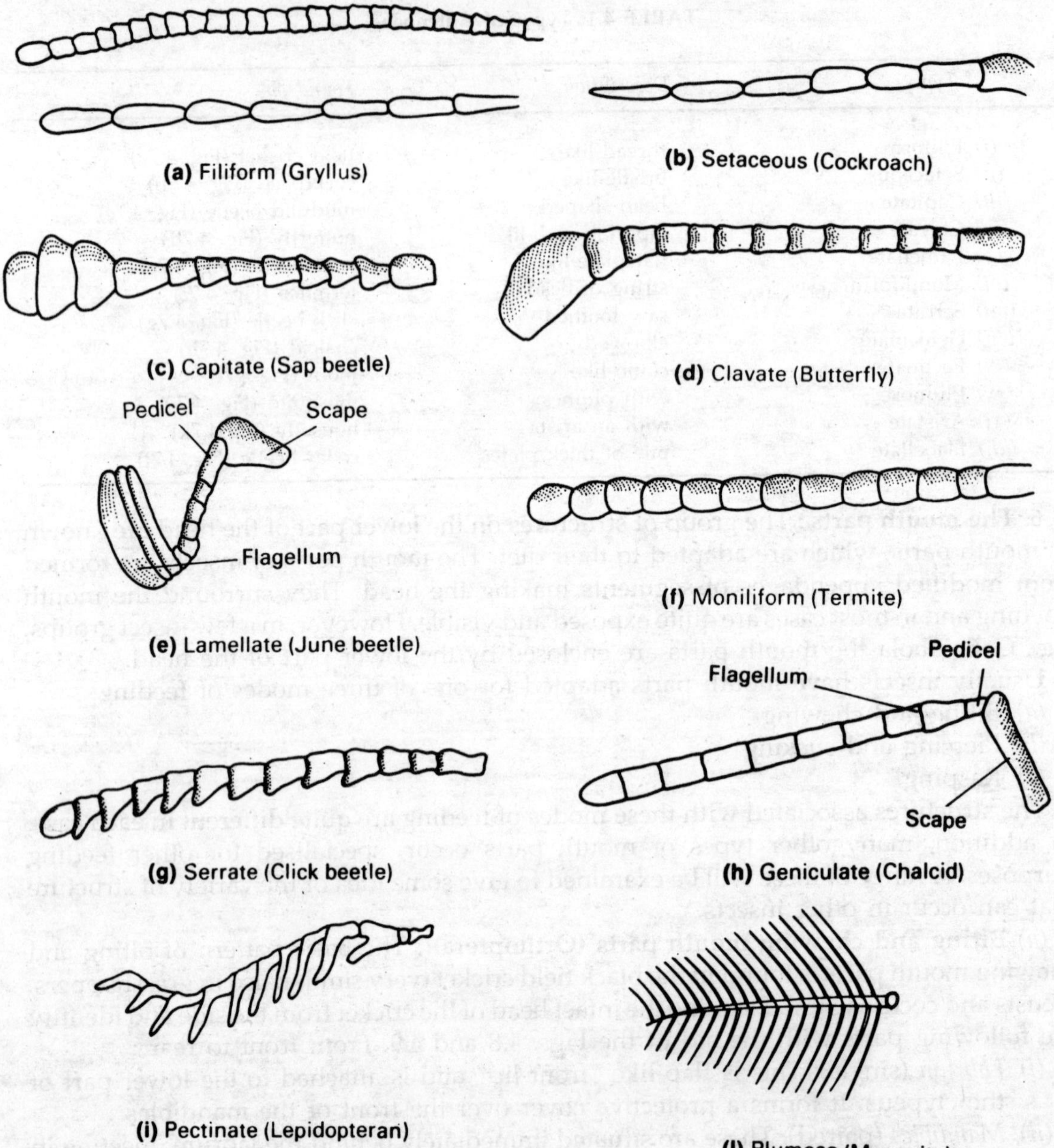

Fig. 4.7: Types of insect antennae.

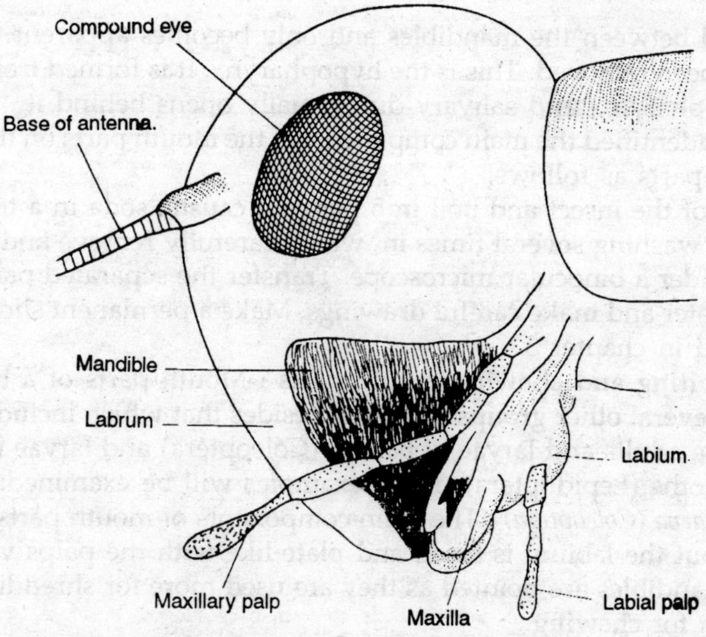

Fig. 4.8: Side view of head to show mouth parts in situ

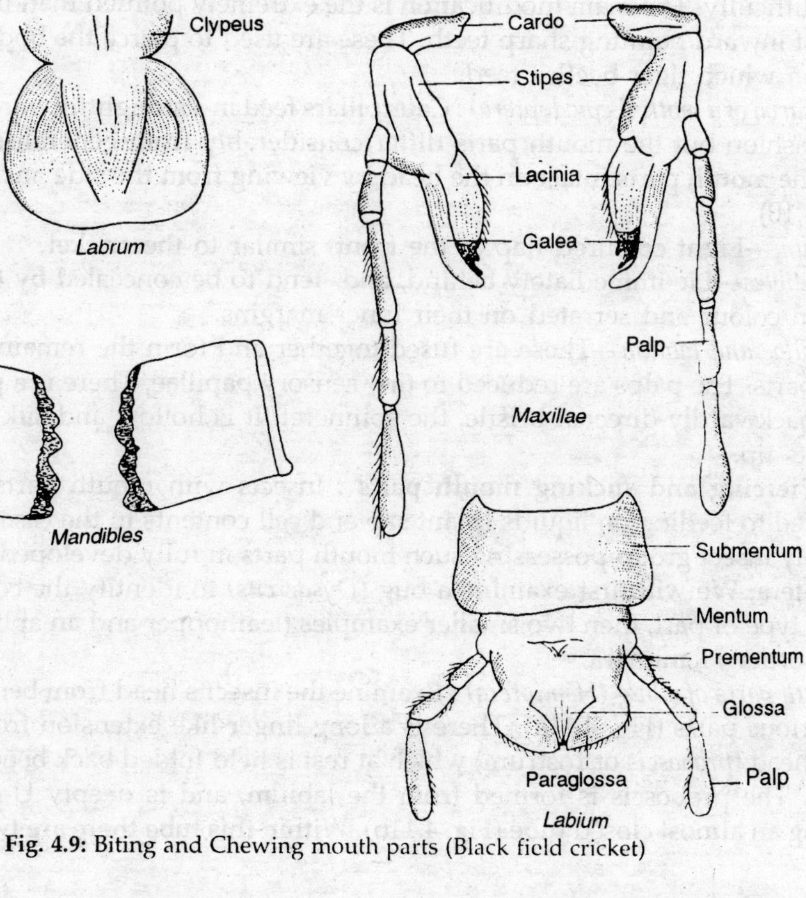

Fig. 4.9: Biting and Chewing mouth parts (Black field cricket)

is concealed between the mandibles and only becomes apparent when all other parts have been removed. This is the hypopharynx. It is formed from the first part of the gut (pharynx) and salivary duct usually opens behind it.

After you have identified the main components of the mouth parts on the intact insect, study the mouth parts as follows:

· Take the head of the insect and boil in 5 per cent caustic soda in a test tube for 15-20 minutes. After washing several times in water, carefully remove and see the mouth parts in water, under a binocular microscope. Transfer the separated parts into a clean petri dish with water and make careful drawings. Make a permanent slide of the mouth parts as discussed in chapter 3.

(b) Modified biting and chewing mouth parts : Mouth parts of a biting/chewing pattern occur in several other groups of insects besides that which includes the crickets. The main ones are adults and larvae of beetles (Coleoptera) and larvae (caterpillars) of butterflies and moths (Lepidoptera). Three examples will be examined:

(i) Grass grub larva (Coleoptera) : The main components of mouth parts can be readily identified but the labium is small and plate-like with the palps very reduced in size. The mandibles are pointed as they are used more for shredding plant tissue (roots) than for chewing.

(ii) Tiger beetle adult (Coleoptera) : The main components can again be identified without difficulty. The main modification is the extremely pointed mandibles with a series of inward pointing sharp teeth. These are used to pierce the body of other insects on which tiger beetles feed.

(iii) Larva of a moth (Lepiodoptera) : Caterpillars feed in a straightforward biting/chewing fashion but the mouth parts differ considerably from generalised plan. Examine the mouth parts intact on the head by viewing from the side and underneath (Fig. 4.10).

Labrum—Light coloured flap at the front, similar to the cricket.

Mandibles—Lie immediately behind, and tend to be concealed by the labrum. Black in colour and serrated on their inner margins.

Maxillae and labium—These are fused together and form the remainder of the mouth parts. The palps are reduced to tiny sensory papillae. There is a prominent central backwardly-directed bristle, the spinneret. It is hollow and silk is extruded from its tip.

(c) Piercing and sucking mouth parts : Insects with mouth parts of this type are restricted to feeding on liquids (plant sap and cell contents in the case of plant feeders). The only insect group possessing such mouth parts in fully developed form is the order Hemiptera. We will first examine a bug *(Dysdercus)* to identify the components of this mouth type of part, then two smaller examples (leafhopper and an aphid)—all members of the order Hemiptera.

Mouth parts of a bug (Hemiptera) : Examine the insect's head from beneath and identify the various parts (Fig. 4.11a). There is a long finger-like extension from the lower part of the head *(proboscis* or rostrum) which at rest is held folded back beneath the head and thorax. The proboscis is formed from the labium, and is deeply U-shaped in section forming an almost closed tube (Fig. 4.11b). Within this tube there are two pairs of bristle-

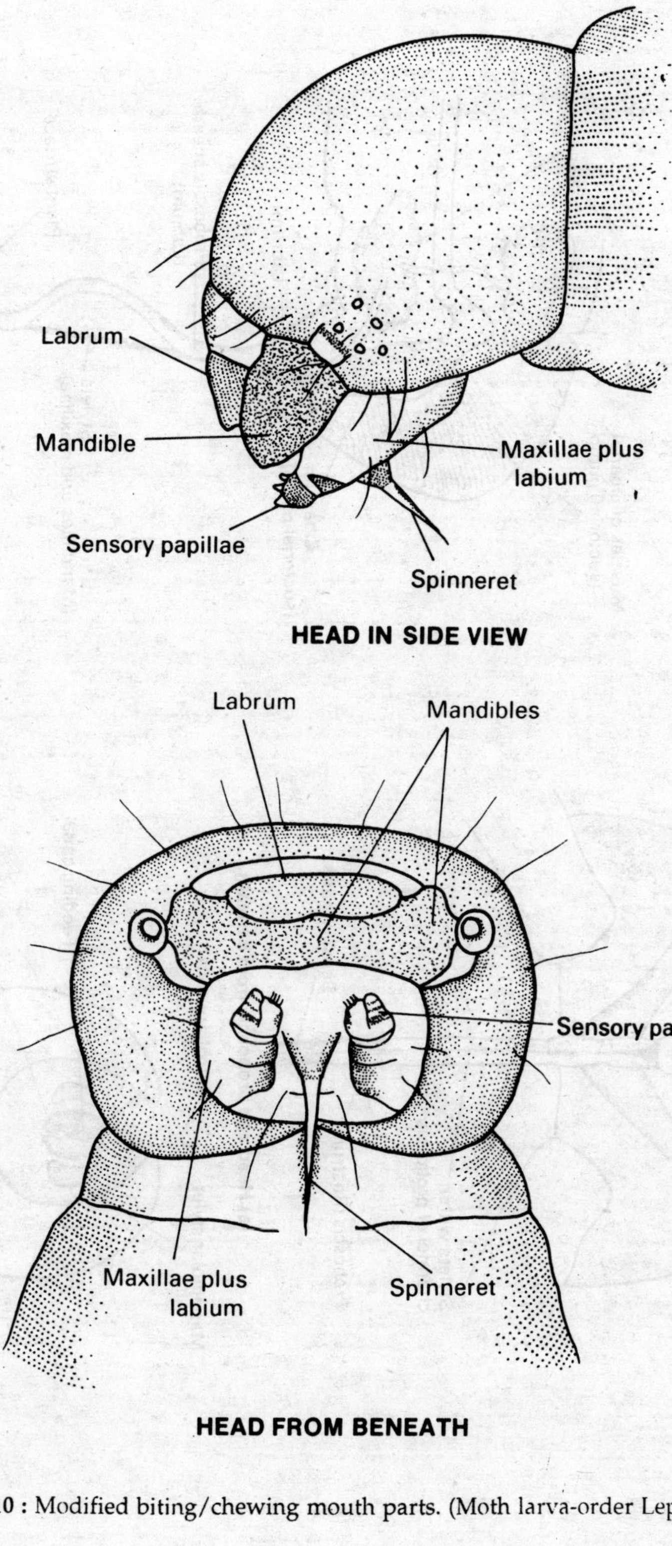

Fig. 4.10 : Modified biting/chewing mouth parts. (Moth larva-order Lepiodoptera)

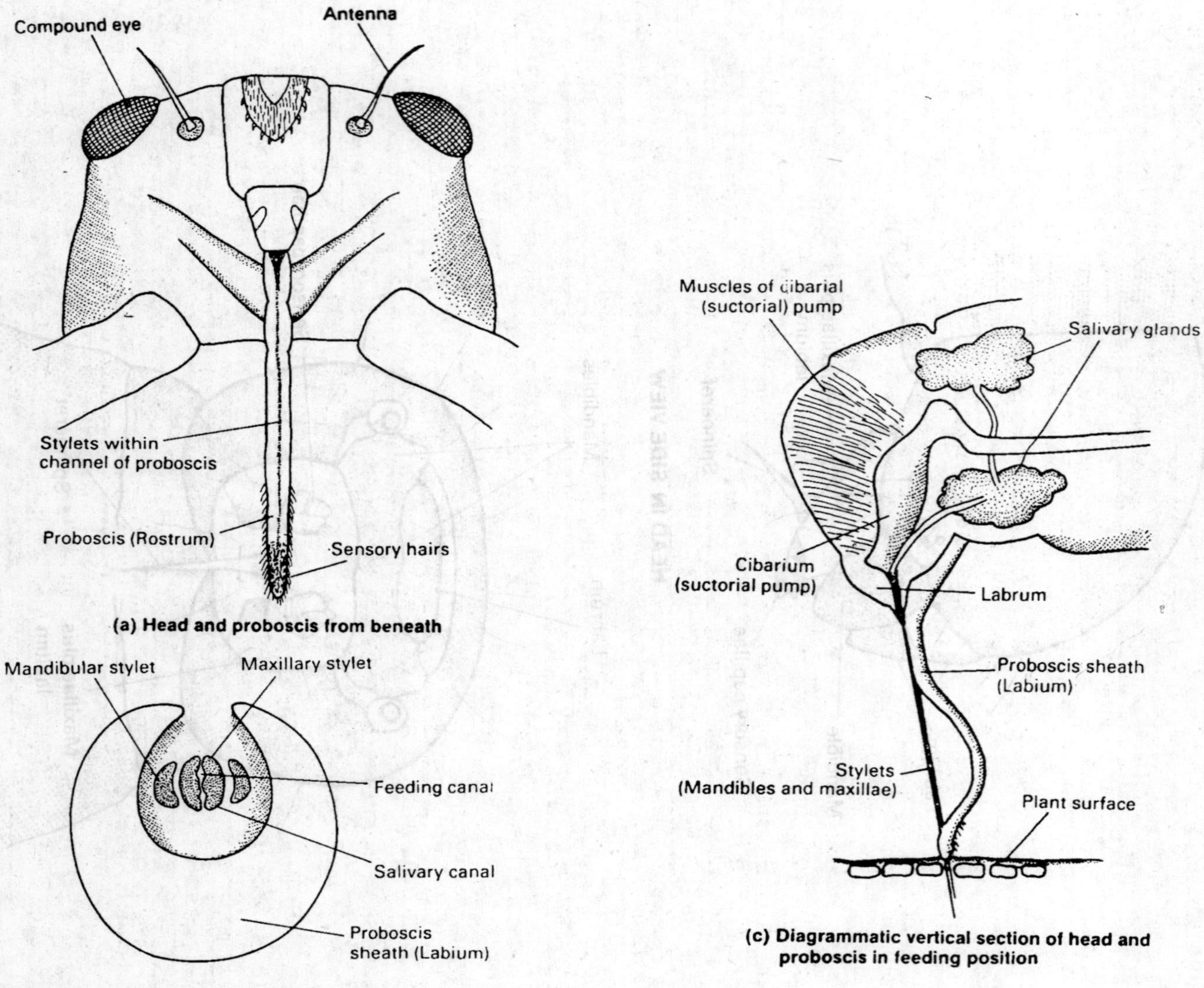

(a) Head and proboscis from beneath

(b) Diagrammatic cross-section of proboscis and stylets

(c) Diagrammatic vertical section of head and proboscis in feeding position

Fig. 4.11: Piercing and sucking mouth parts (Adult Dysdercus order Hemiptera)

like *stylets* formed from the highly modified maxillae and mandibles. Palps are absent. In feeding, the stylets pierce the plant tissue, the proboscis sheath acting as a guide and folding out of the way as the stylets penetrate (Fig. 4.11c). The inner *maxillary stylets* bear a pair of fine grooves along their inner surfaces. These two stylets are normally locked firmly together so that the opposing grooves form two fine tubes. Saliva is pumped down the posterior of these tubes and plant sap is taken up by the other (Fig. 4.11b). Complete separation of the stylets is often not possible without breaking them, so that although there are four stylets present, one may not be able to count them all. The outer *mandibular stylets* assist in penetration of plant tissue but are not otherwise actively involved in the feeding process.

To free the stylets from the proboscis sheath in your specimen, carefully insert the tip of a sharp needle at the base of the proboscis and slowly work them loose towards the tip. Although palps are absent from piercing/sucking mouth parts note that the rostrum bears a number of *sensory hairs* towards its tip.

Mouth parts of a leafhopper and of an aphid (Hemiptera): Most insects that are important as plant pests and which possess piercing/sucking mouth parts are much smaller than bugs, so it is appropriate to examine two examples. Examine the mouth parts of the leafhopper and the aphid and identify the component parts as for the bug. Keep these small, rather soft specimens, in alcohol in a watch glass, otherwise they will collapse and shrivel. In all members of the Hemiptera the basic form of the mouth parts is identical to that of the bug. They differ only in size and to some extent in proportions.

Besides these plant feeding insects, other insects feeding on animals' fluid have piercing and sucking type of mouth parts. We shall discuss only two examples—mosquitoes *(Diptera)* and fleas *(Siphonaptera)*.

Mouth parts of a female mosquito (Diptera): The labrum, mandibles and maxillae (sometimes hypopharynx also) are slender and long and fit together to form a delicate hollow needle. The labium forms a stout sheath that holds this needle rigid (Fig. 4.12). The entire structure is called a beak. During feeding, the insect presses the entire beak against the host and inserts the needle into the host tissues. This is how the host fluid is sucked through the needle into the oesophagus of the insect.

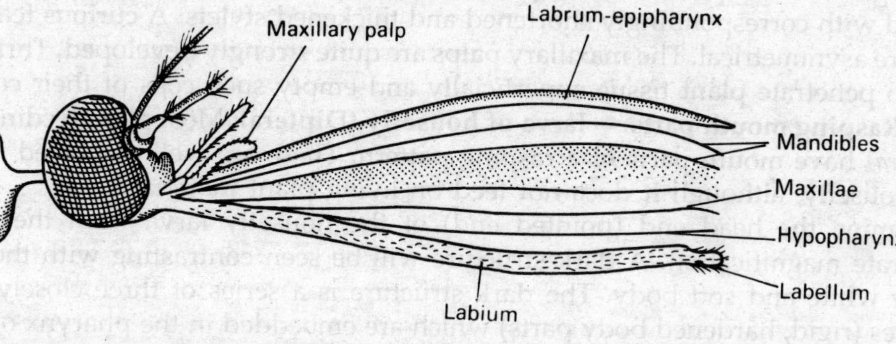

Fig. 4.12: Piercing and sucking mouth parts of female anopheles (Mosquito)

Mouth parts of a flea (Siphonaptera): Though mouth parts of a flea are adapted for piercing and sucking type of feeding, the most specialised parts are the *mandibles*. These are rather broad plates, serrated along the distal two-thirds of their length having a groove on the inner side, forming together a tube or channel through which saliva is ejected. The proximal one-third of the mandibles is in contact with the short hypopharynx. The *labrum-epipharynx* is a long slender organ which is ventrally grooved and lies in close apposition to the mandibles, thus forming an afferent channel through which blood is sucked up (Fig. 4.13).

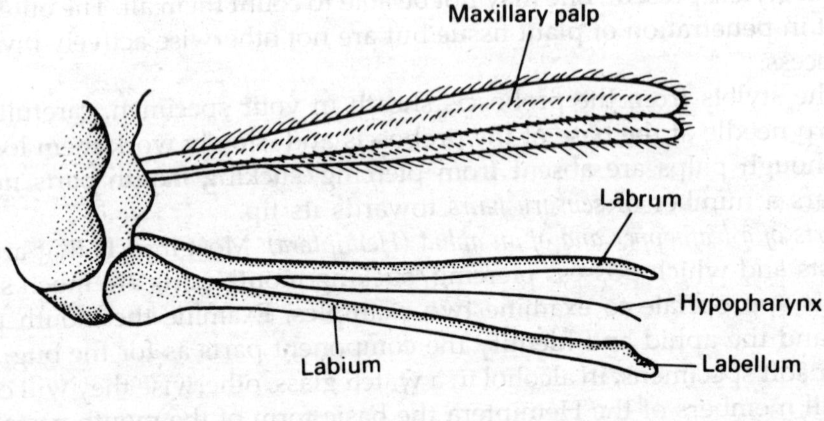

Fig. 4.13 : Piercing and blood/sucking mouth parts of Glossina palpalis

Each *maxilla* consists of a single lobe on either side of the mouth, and a four-jointed palpus; they are not cutting organs and do not enter the host tissues. The *labium* is made up of a basal plate or mentum from which arises a pair of five-jointed labial palps. When feeding the flea punctures the skin by mandibles and labrum-epipharynx. The muscles of the salivary pump inject saliva into the wound so formed and the afferent channel sucks up the blood into the pharynx.

(d) Mouth parts similar to the piercing and sucking pattern: Only one other group of plant feeding insects, i.e., **thrips** *(Thysanoptera)* possesses mouth parts similar to the piercing/sucking type of the Hemiptera. In this case the proboscis is very short and cone-shaped with correspondingly shortened and thickened stylets. A curious feature is that they are asymmetrical. The maxillary palps are quite strongly developed. Thrips are only able to penetrate plant tissue superficially and empty such cells of their contents.

(e) Rasping mouth parts — larva of housefly (Diptera): Most plant feeding fly larvae *(Diptera)* have mouth parts of a rasping pattern. These are well displayed in the larva of a housefly, although it does not feed on living plant material.

Examine the head end (pointed end) of the housefly larva from the side under moderate magnification. A dark structure will be seen contrasting with the otherwise totally white and soft body. The dark structure is a series of three closely associated sclerites (rigid, hardened body parts) which are embedded in the pharynx of the insect. The tip of the front sclerite is in the form of a pair of sharp recurved claws (mouth hooks*)

*The mouth hooks are clearly paired in the larvae of most fly species but in the larva of housefly there appears to be only a single hook.

which can be extruded from the mouth (Fig. 4.14a). By moving the head around it, the larva can use the mouth hooks to rasp away at food material. The resultant fluid and shredded food particles are then ingested.

The structure of rasping type of mouth parts cannot be related to other insect mouth part types and probably have evolved quite independently. The full structure can only be seen in a cleared slide preparation (Fig. 4.14b).

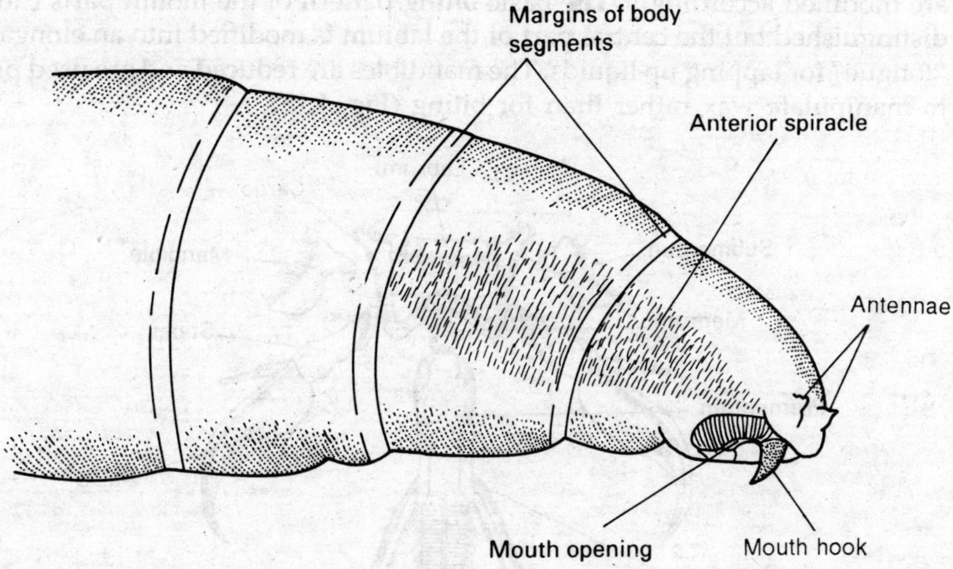

(a) Head of intact larva-side view

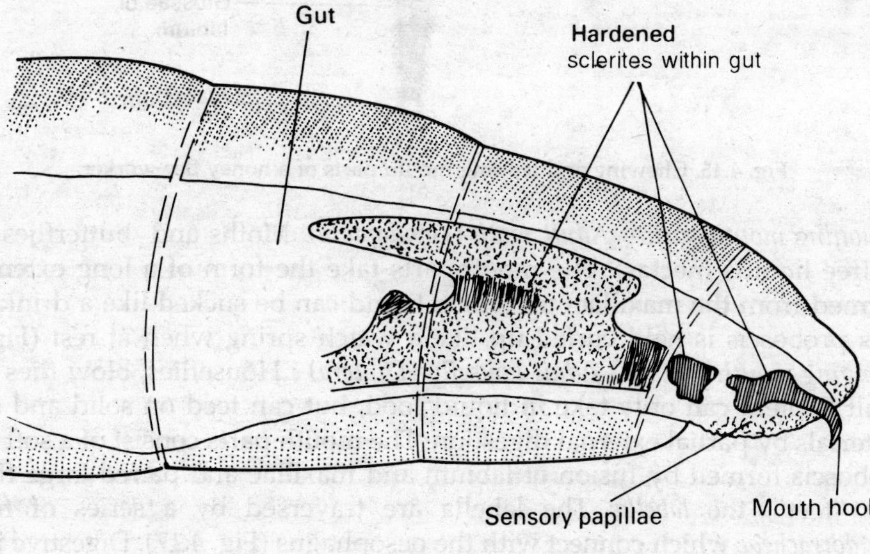

(b) Slide preparation of head to show internal structure

Fig. 4.14: Rasping mouth parts-larva of housefly (order Diptera)

(f) Other types of Mouth parts: Three other insect mouth part types will be discussed. These are of insects which are *not* plant feeding; at least not plant damaging, though feeding on nectar and pollen may take place.

(i) Lapping mouth parts—Honeybee worker (Hymenoptera): Honey bees can deal with semi-solid food (honey) but also can lap up liquids (nectar). They are also required to manipulate solid material in construction of the combs (wax). The mouth parts are modified accordingly. The basic biting pattern of the mouth parts can still be distinguished but the central part of the labium is modified into an elongate hairy "tongue" for lapping up liquids. The mandibles are reduced and are used primarily to manipulate wax rather than for biting (Fig. 4.15).

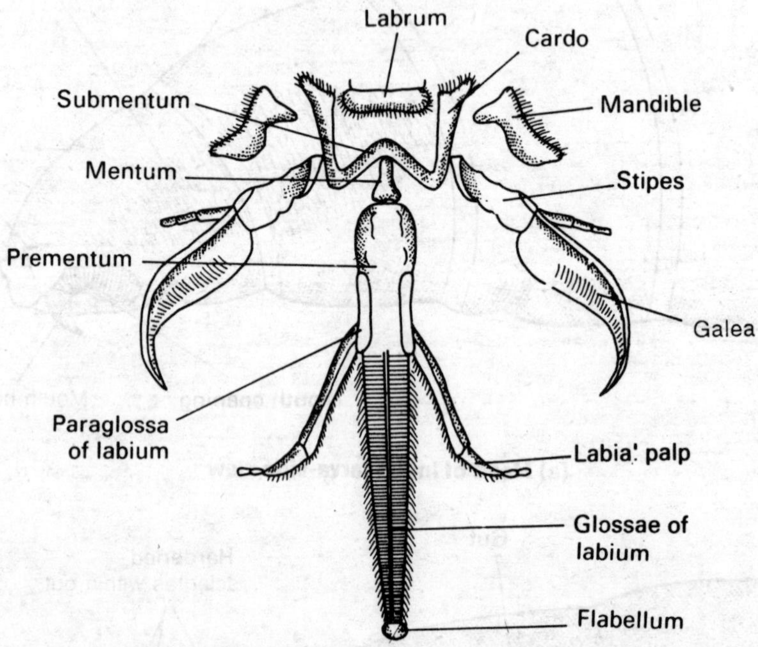

Fig. 4.15. Chewing and lapping mouth parts of a honey Bee worker.

(ii) Siphoning mouth parts—Adult moth (Lepidoptera): Moths and butterflies feed only on free liquids (nectar). The mouthparts take the form of a long extensible tube (formed from the maxillae) up which liquid can be sucked like a drinking straw. This proboscis is held curled up like a watch spring when at rest (Fig. 4.16).

(iii) Sponging mouth parts — Adult housefly (Diptera) : Houseflies, blow flies and other adult Diptera can only take in liquid food, but can feed on solid and semi-solid materials by partial external digestion. The mouth parts consist of a soft extensible proboscis formed by fusion of labium and maxillae and paired large fleshy pad-like "feet", the *labella.* The labella are traversed by a series of fine tubes, *pseudotracheae* which connect with the oesophagus (Fig. 4.17). Digestive juices pass down these tubes and liquify the food, which can then be sucked up in the food channel formed by the labrum-epipharynx and the hypopharynx, much as in the mosquito, but without these two structures being developed for piercing.

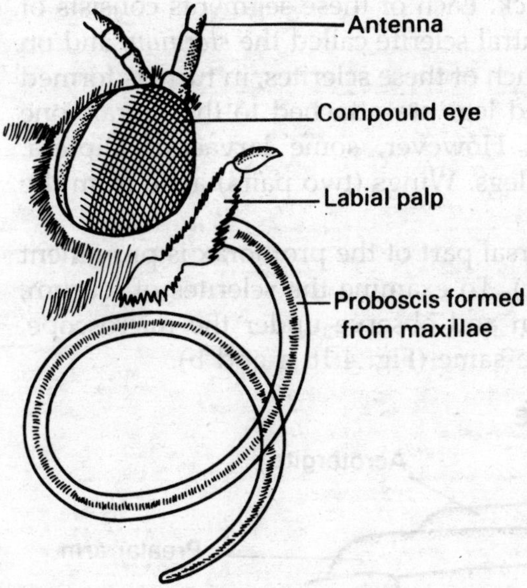

Fig. 4.16. Siphoning mouth parts of Butterfly.

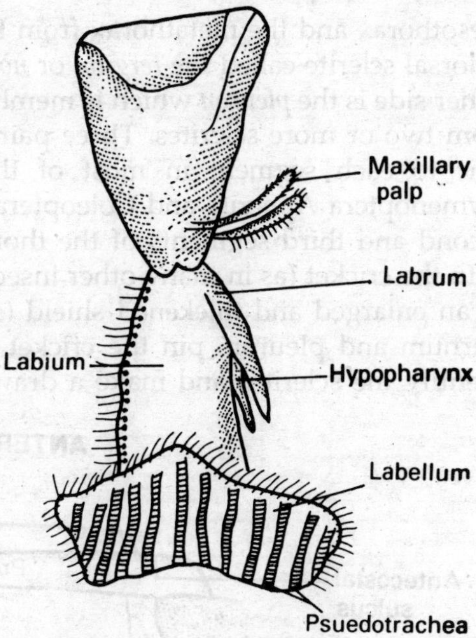

Fig. 4.17. Sponging mouth parts of *Musca domestica* (Housefly)

Recognition of the types of Mouth Parts

The importance of having an understanding of different insect mouth part types is that injury to plants caused by each type is quite different. Thus for instance, insects with piercing/sucking mouth parts are quite incapable of removing solid plant tissue (and are thus incapable of producing holes in leaves or other plant parts), but they do inject saliva when they feed and this can have serious effects on plant growth.

Individual species of plant feeding insects usually attack only one particular part of the plant, e.g., leaves or fruit or stems or roots. Therefore in order to answer the following section, you will have to find out what part of the plant the particular pest attacks, as well as the type of mouth parts it possesses.

Name of insect	*Mouth part type*	*Brief description of plant injury*
1.		
2.		
3.		

2. THE THORAX

1. Structure: The thorax in all insects consists of three segments, the prothorax , the mesothorax and the metathorax from front to back. Each of these segments consists of a dorsal sclerite called the *tergum* or *notum*, a ventral sclerite called the *sternum*, and on either side is the *pleuron* which is membranous. Each of these sclerites, in turn, is formed from two or more sclerites. Three pairs of jointed legs are attached to the thorax, one pair to each segment in most of the insects. However, some larvae of Diptera, Hymenoptera Apocrita and Coleoptera have no legs. Wings (two pairs) arise from the second and third segments of the thorax.

In the cricket (as in many other insects) the dorsal part of the prothorax is prominent as an enlarged and thickened shield (*see* Fig. 4.1). To examine the sclerites of tergum, sternum and pleuron, pin the cricket in position and observe under the microscope. Identify the sclerites and make a drawing of the same (Fig. 4.18 a and b).

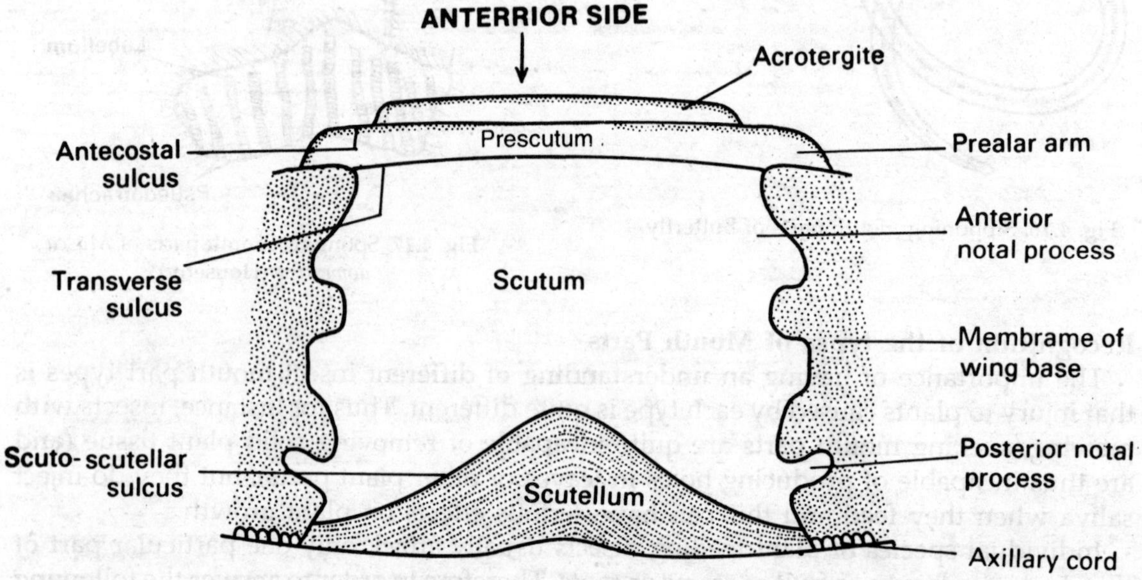

Fig. 4.18 (a) : Diagrammatic representation of notum of a wing-bearing segment.

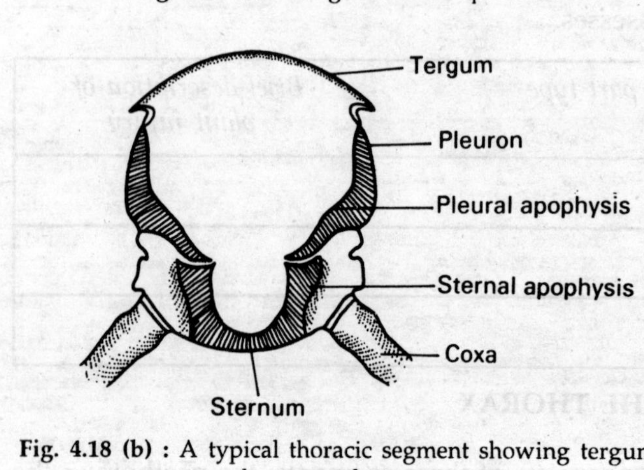

Fig. 4.18 (b) : A typical thoracic segment showing tergum, pleuron and sternum

2. The legs: Insect legs have great structural variations adapted to life habits of insects. However, the basic plan of all the legs is similar. Each leg consists of a coxa (articulating with the pleuron and sternum by a membrane), trochanter, femur, tibia, tarsus, pretarsus and two claws and a pad (called arolium).

Remove one of the metathoracic legs of cricket, examine, draw and label the parts (Fig. 4.19a). Study the legs of following insects to characterise different modifications in insect legs:

(i) Metathoracic leg of black field cricket—*ambulatorial leg* (for walking) (Fig. 4.19a).

(ii) Mesothoracic leg of *Periplaneta americana*—*cursorial leg* (for running) (Fig. 4.19b).

(iii) Metathoracic leg of *Apis indica*—with pollen basket (for collection of pollen) (Fig. 4.19c).

(iv) Prothoracic leg of *Mantis religiosa*—*raptorial leg* (for catching the prey) (Fig. 4.19d).

(v) Prothoracic leg of *Gryllotalpa africana*—*fossorial leg* (for digging in the soil) (Fig. 4.19e).

(vi) Prothoracic leg of *Haematopinus suis*—*scansorial leg* (for clinging) (Fig. 4.19f).

(vii) Prothoracic leg of grasshopper—with tympanum (for hearing) (Fig. 4.19g).

(viii) Metathoracic leg of a tettigoniid—with stridulatory organ (for sound production) (Fig. 4.19h).

(ix) Metathoracic leg of water scavenger beetle—*natatorial leg* (for swimming) (Fig. 4.19i).

(x) Metathoracic leg of grasshopper—*saltatorial leg* (for jumping) (Fig. 4.19j).

(xi) Prothoracic leg of male *Dytiscus* sp.—*clasping leg* (Fig. 4.19k).

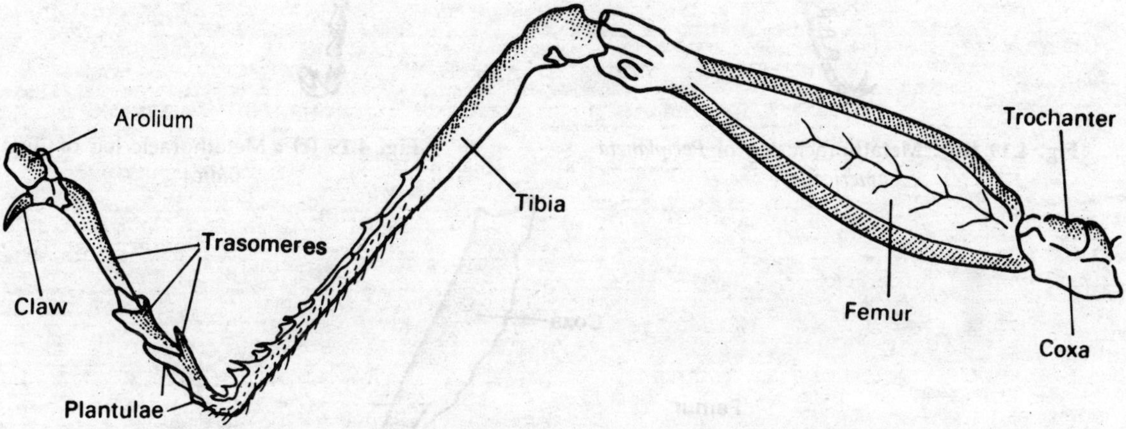

Fig. 4.19 (a) : Metathoracic leg of black field cricket.

3. **The wings:** Wings when present arise from the second and third segments of the thorax as outgrowths of the body wall. There are typically two pairs but in a few groups, such as the true flies *(Diptera)*, they are reduced to a single pair. Many insects are of course wingless (apterygota). The wings themselves are mostly membranous but in beetles *(Coleoptera)* the forewings take the form of hard horny cases which are not involved in flight. In all wings except the smallest of insects the membranous part of the wing is supported by a series of thickened ribs, called *veins.* The pattern of arrangement of these veins, their branches and cross connections, varies greatly in different insect groups and is an important feature in classification. The basic plan of wing venation and a range of insect wing types is shown in Figs. 4.20 and 4.21.

4. **Features found in only a few insects:** *File* and *scraper*: Sound producing mechanism on the undersides of the forewings and forelegs of some males (Fig. 4.22 a and b). Sound is used as a mating call.

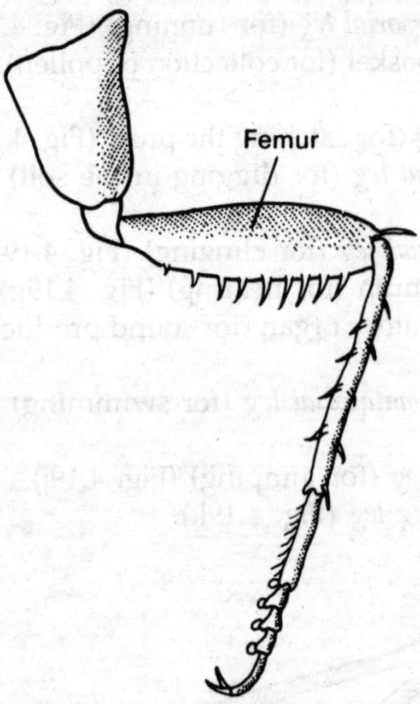

Fig. 4.19 (b) : Metathoracic leg of *Periplaneta americana*

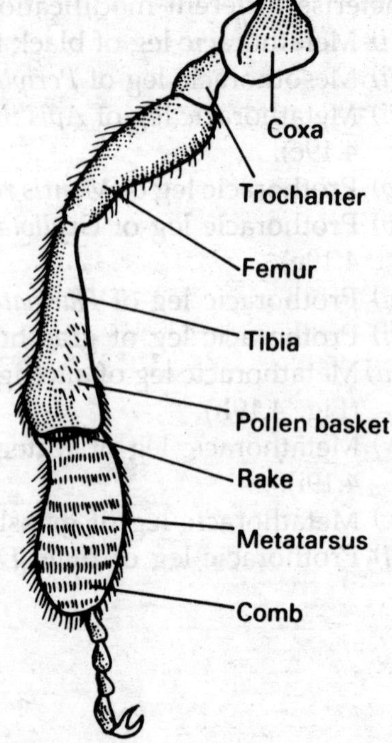

Fig. 4.19 (c) : Metathoracic leg of *Apis indica*

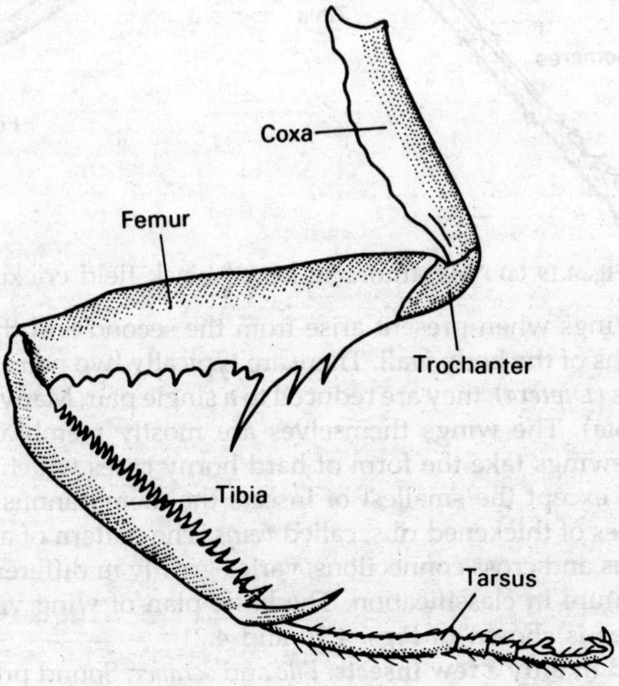

Fig. 4.19 (d) : Prothoracic leg of *Mantis religiosa*

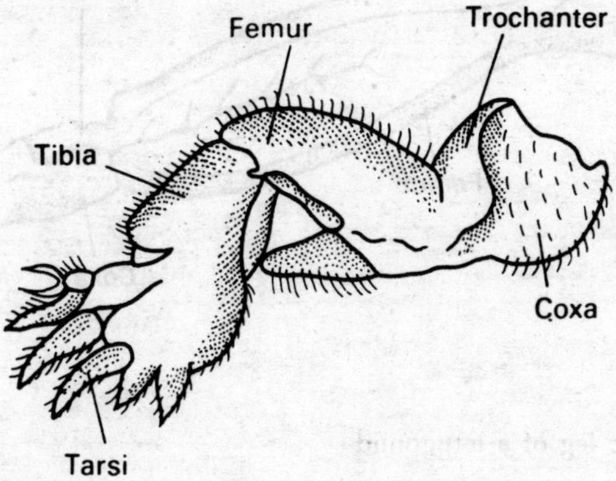

Fig. 4.19 (e) : Prothoracic leg of *Gryllotalpa africana*

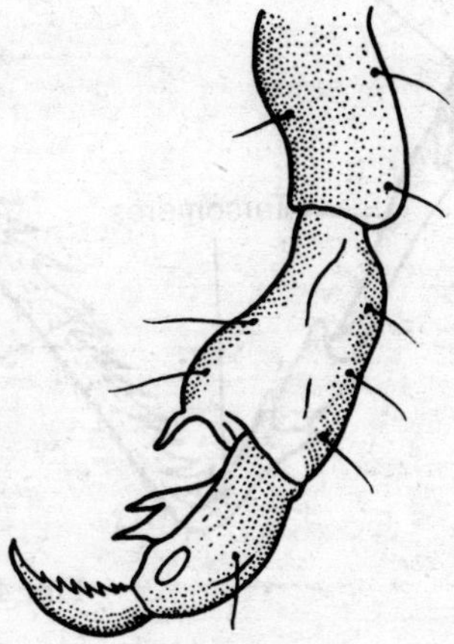

Fig. 4.19 (f) : Prothoracic leg of *Haematopinus suis*

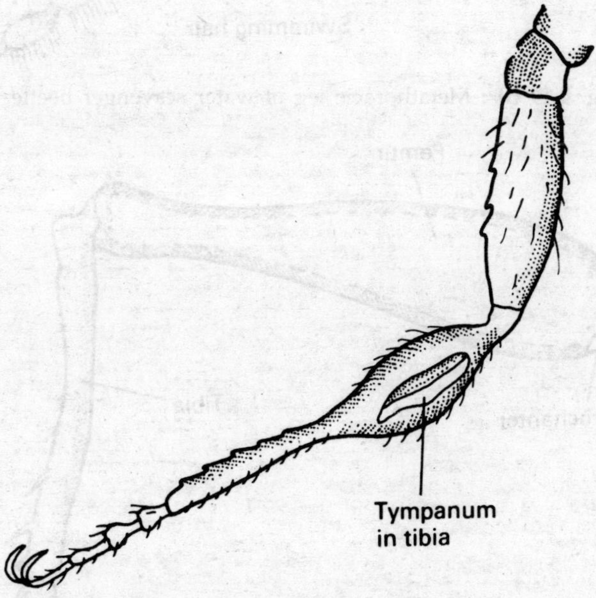

Fig. 4.19 (g) : Prothoracic leg of grasshopper

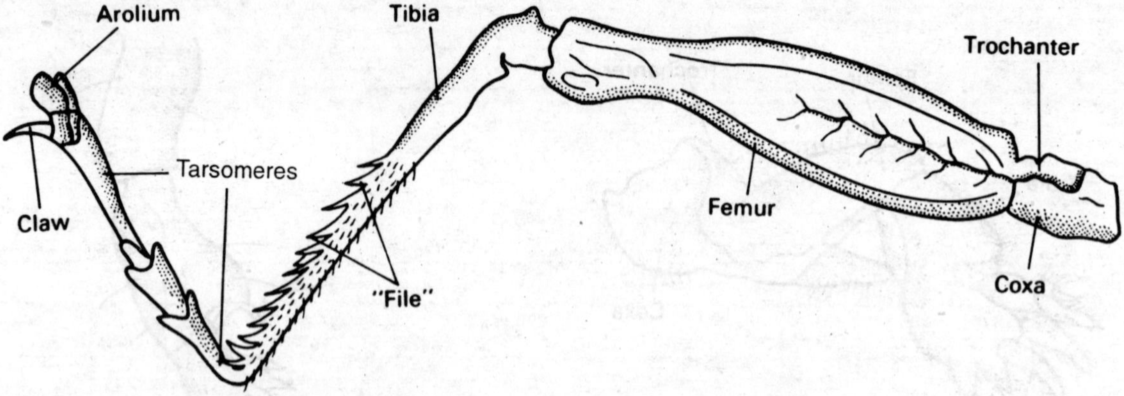

Fig. 4.19 (h) : Metathoracic leg of a tettigoniid

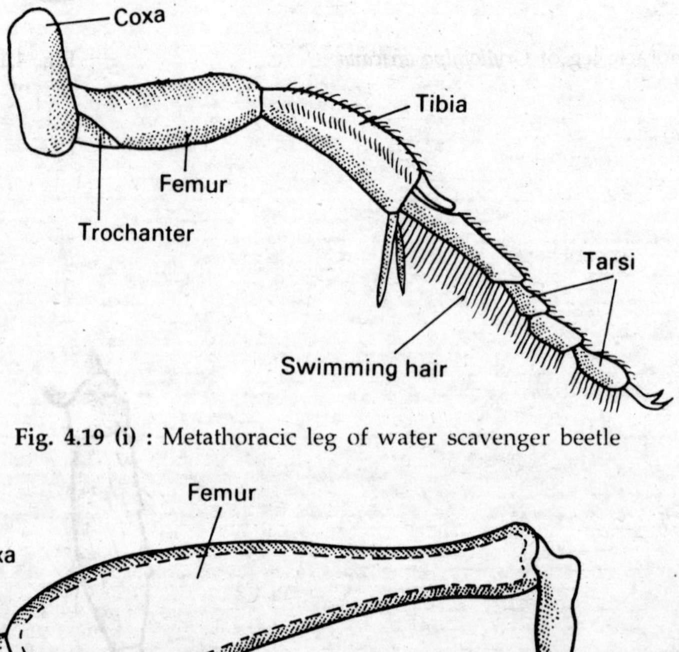

Fig. 4.19 (i) : Metathoracic leg of water scavenger beetle

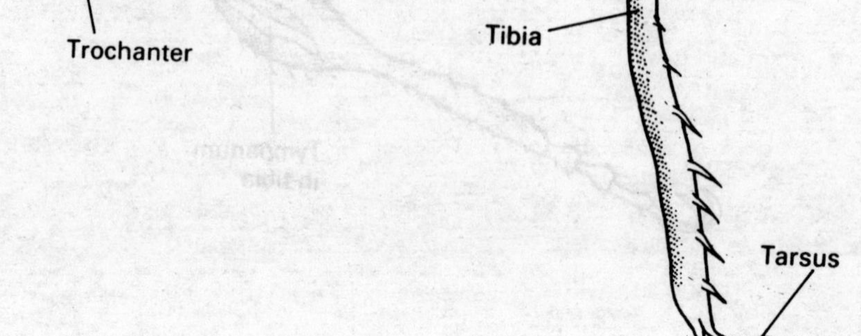

Fig. 4.19 (j) : Metathoracic leg of grasshopper

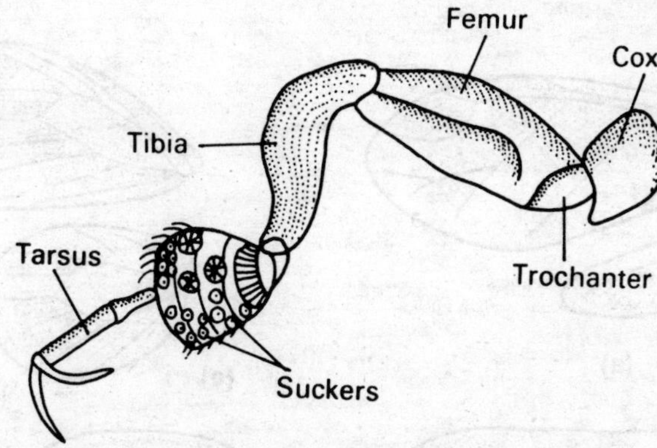

Fig. 4.19 (k) : Prothoracic leg of male *Dytiscus*.

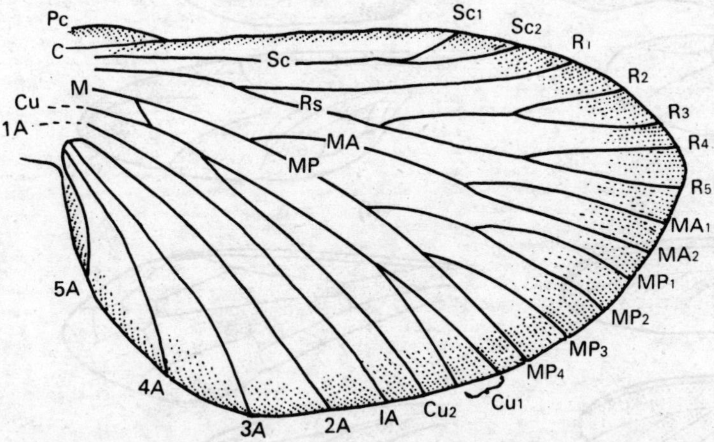

Fig. 4.20 (a) : Arche type venation of an insect wing with veins named according to the Comstock-Needham system.

A = Anal veins
Ax = Axillaries
C = Costa
Sc = Subcosta
Cu = Cubitus
R = Radial cross vein
Rs = Radial sector
S = Suctorial cross vein
tg = Tegula
M = Media
m, m´ = Median plates
MA = Media anterior
MP = Media posterior
J = Jugal veins
V = Vannal veins
jf = Jugal fold
Vf = Vannal fold
HP = Humeral plate
PCu = Post cubitus

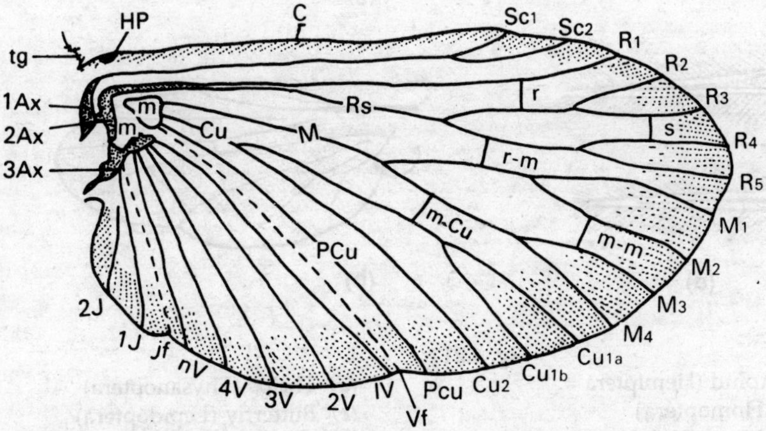

Fig. 4.20 (b) : The usual wing veins and axillaries.

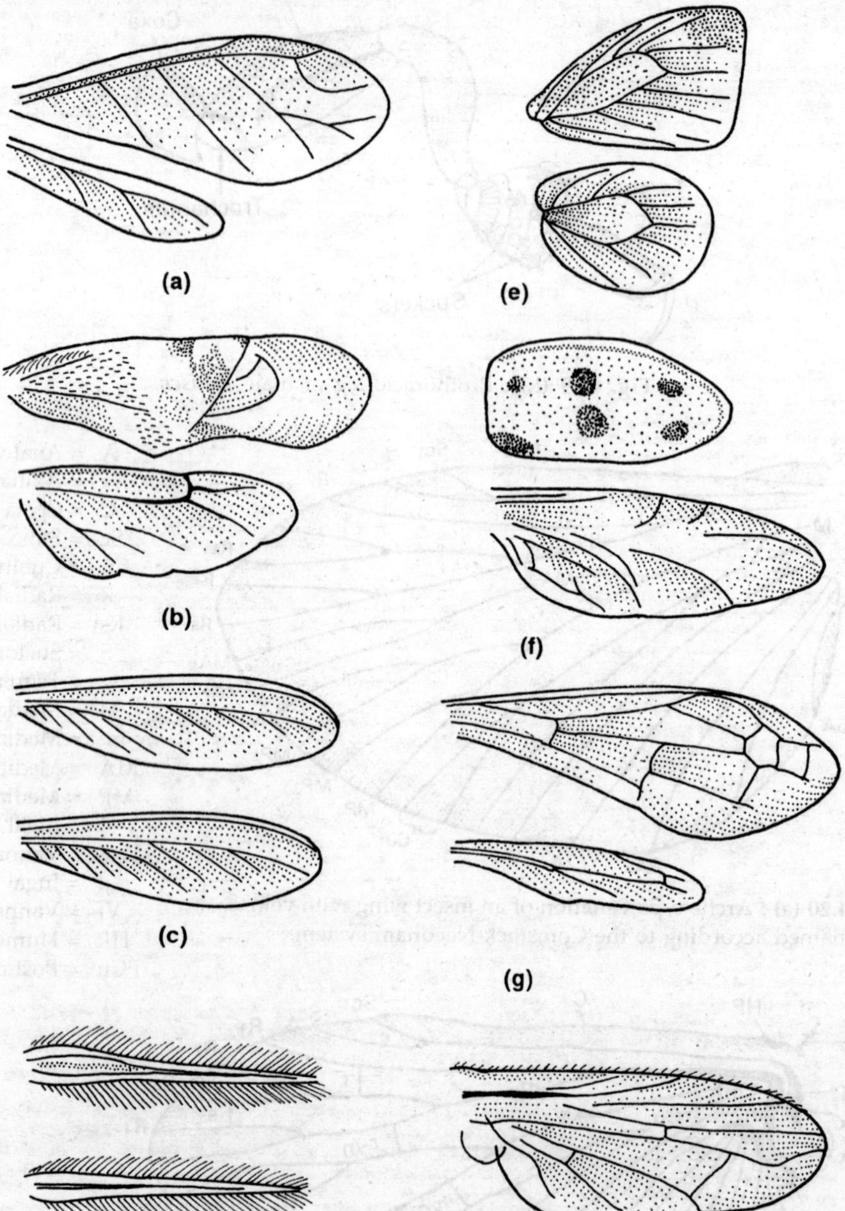

Fig. 4.21 :

(*a*) Aphid (Hemiptera = Homoptera)
(*b*) Mirid (Hemiptera — + Heteroptera)
(*c*) Termite (Isoptera)
(*d*) Thrips (Thysanoptera)
(*e*) Butterfly (Lepidoptera)
(*f*) Ladybird beetle (Coleoptera)
(*g*) German was (Hymenoptera)
(*h*) Housefly (Diptera)

Sound receiving organ—ear (tympanum) present as a sunken whitish patch towards the top of the tibia of the forleg (Fig. 4.19g).

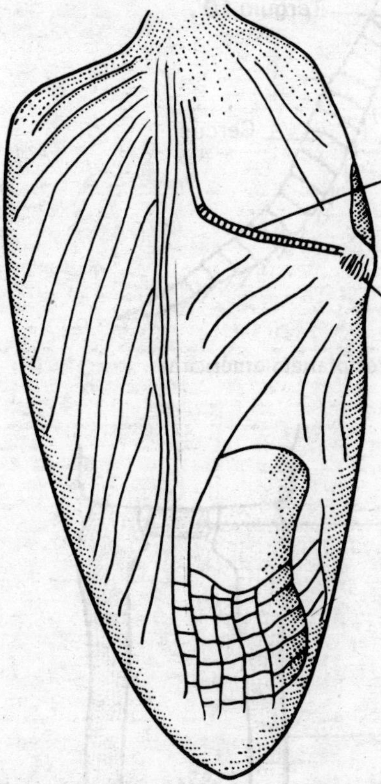

Fig. 4.22 (a) : Underside of forewing of male cricket.

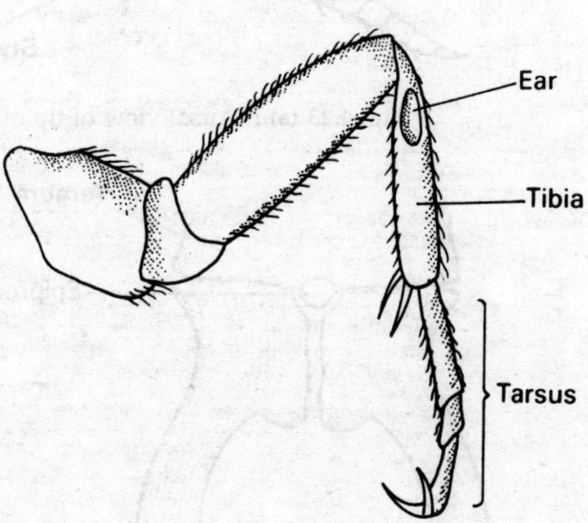

Fig. 4.22 (b) : Foreleg of cricket.

3. THE ABDOMEN

The abdomen is composed of eleven segments in cricket but only nine can be counted as the last two are reduced and retracted within the others (*see* Fig. 4.1). Each abdominal segment is covered with hard dorsal (tergum) and ventral plates (sternum) separated by a membranous area at each lateral side. The dorsal and ventral plates are joined together by membranous intersegmental membranes. There are small dark openings, the *spiracles* set into the soft membranous area along each side of the abdomen. There is one pair to each segment. Two pairs of spiracles are also present on the thorax but are difficult to see.

(a) Cerci: These are a pair of tapering appendages arising from the tip of the abdomen (segment 11). In *Periplaneta americana* (Dictyoptera), cerci are simple, jointed and clothed in fine hair (touch sensitive) while in Dermaptera, these are sclerotized and look-like forceps (Fig. 4.23 a and b).

(b) Genitalia: External reproductive organs or genitalia are present on the abdomen of many insects. In the female these may take the form of a sword-like projection, the *ovipositor* arising from beneath the eighth and ninth abdominal segments and directed

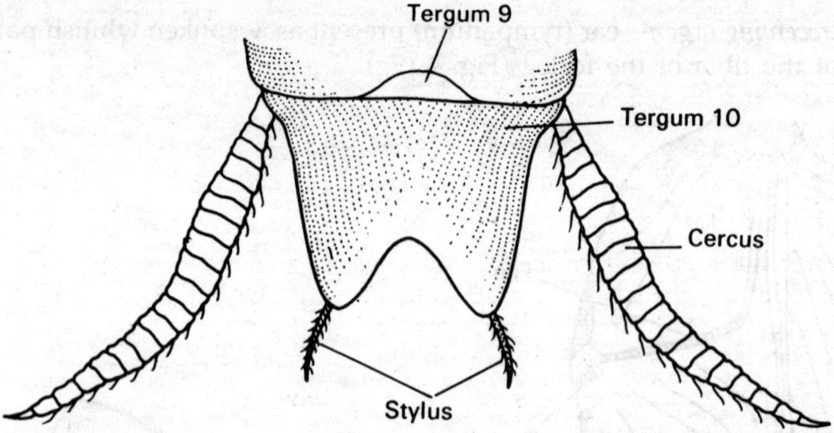

Fig. 4.23 (a) : Dorsal view of tip of abdomen of Periplaneta americana

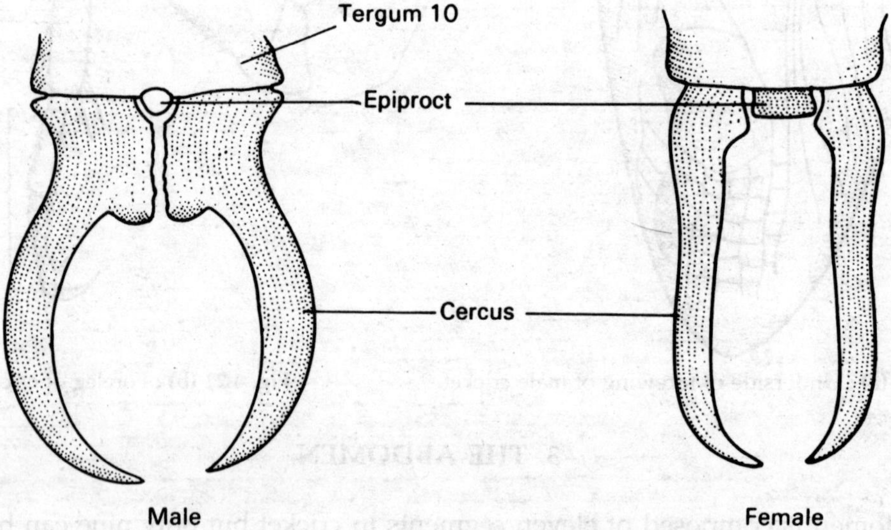

Fig. 4.23 (b) : Male and female cerci of Forficulipa

backwards. The ovipositor consists of up to three pairs of separate shafts or valves (Fig. 4.24a) and is well-developed in grasshoppers, crickets and cicadas. It is used for depositing eggs (in soil) which are squeezed along its length between the shafts. In some insect groups such as Diptera, Lepidoptera and Coleoptera, the hind end of the abdomen is telescopic in form and acts as an ovipositor instead (Fig. 4.24 b).

In higher Hymenoptera such as the wasps and bees the ovipositor has become modified to form a poison-injecting appparatus, the **sting** (Fig. 4.24 c), instead of laying eggs. At rest the sting lies in a pocket within seventh abdominal segement, from which it is exserted when in use. There are three pairs of plate-like sclerites—the quadrate, oblong and triangular plates at the base of the sting which articulate with each other.

In male insects genitalia consist of a pair of *claspers* of varying shape and size with which the abdomen of the female is grasped during mating. A central intromittent organ (*penis* or *aedeagus*) is located between the claspers (Fig. 4.25 a and b).

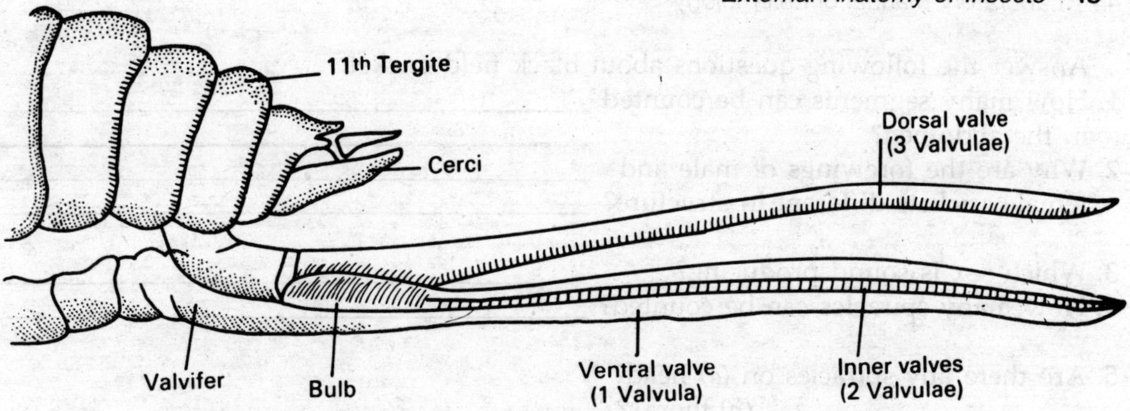

Fig. 4.24 (a) : Lateral view of ovipositor of long-horned grasshopper.

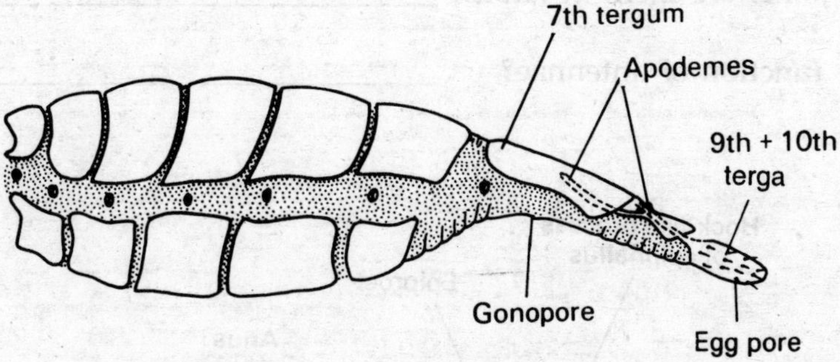

Fig. 4.24 (b) : Telescopic ovipositor of a Lepidoptera

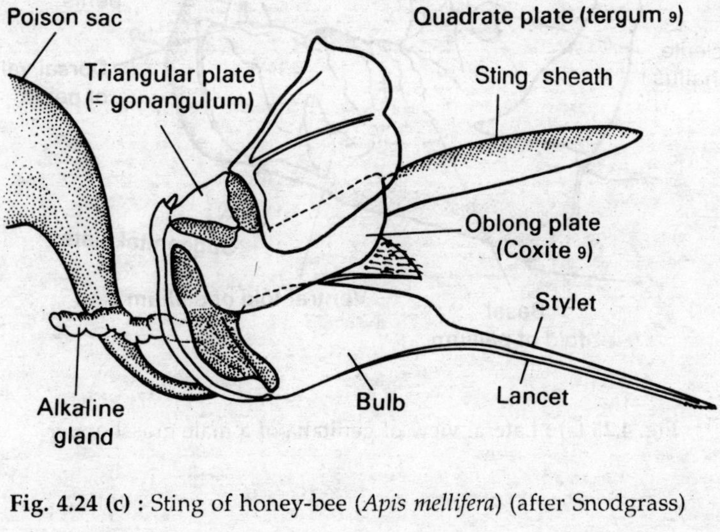

Fig. 4.24 (c) : Sting of honey-bee (*Apis mellifera*) (after Snodgrass)

Answer the following questions about black field cricket:

1. How many segments can be counted in the abdomen? ____________________
2. Why are the forewings of male and female crickets different in structure? ____________________
3. Which sex is sound producing? ____________________
4. How many spiracles can be counted? ____________________
5. Are there any spiracles on *(a)* head? *(b)* thorax? ____________________
6. What is the function of spiracles? ____________________
7. How many ocelli are present? ____________________
8. How many joints are there in tarsus? ____________________
9. What is the function of antennae? ____________________

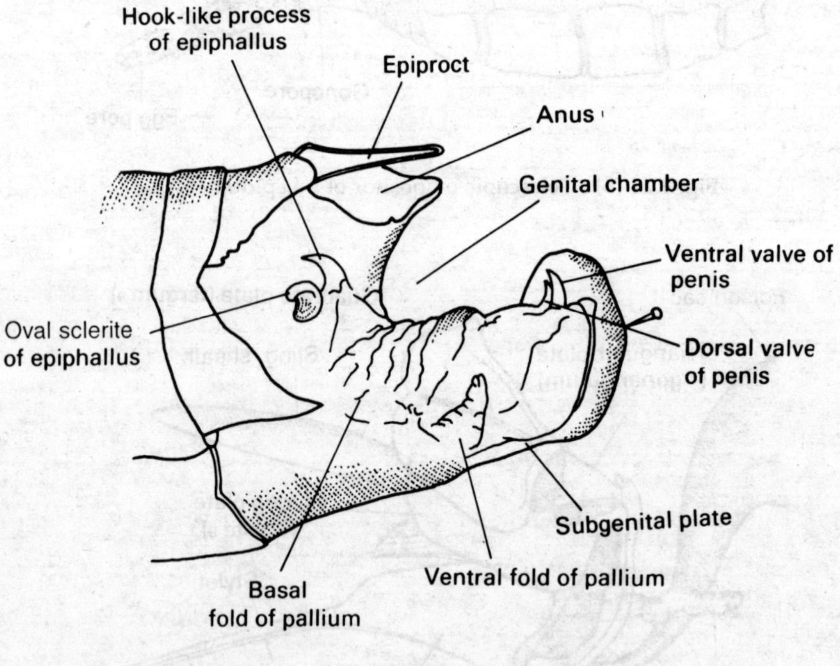

Fig. 4.25 (a) : Lateral view of genitalia of a male grasshopper.

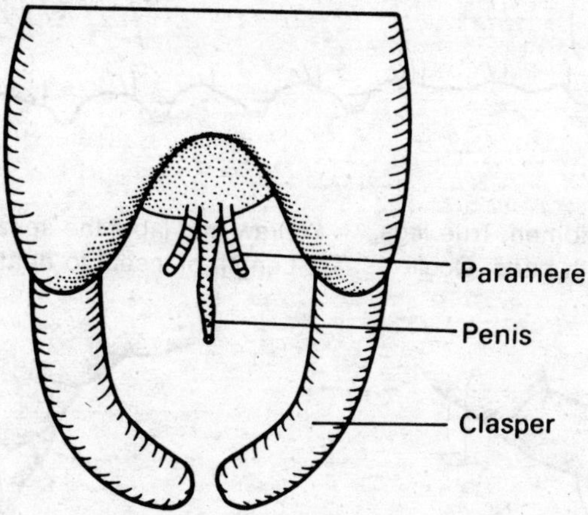

Fig. 4.25 (b) : End-view of male genitalia

LARVA OF A MOTH

Examine the diagram of a moth larva (Fig. 4.26 a) and note the following :

— The head is distinct but there is no marked separation between thorax and abdomen. Segmentation of the thorax and abdomen is however visible.

— Only the head is hard. Rest of the insect is soft, but still possesses a thin exoskeleton which must be shed and renewed when moulting takes place.

— Absence of wings. (Never present in juvenile stages).

The Head

Similar in general shape to that of adult cricket but note that :

— *Antennae* very short and inconspicuous (Fig. 4.26a).

— *Compound eyes* absent (as in all insect larvae).

— Several *ocelli* present in groups on either side of head.

— *Mouth parts* without long, obvious palps.

The Thorax

— Possesses three pairs of jointed legs, but these are much shorter and simpler in structure than those of adult cricket.

— Each leg terminates in a single claw.

The Abdomen

— Count the number of segments.

— On the underside of the abdomen there are several pairs of soft fleshy false legs *(pseudopods)*. They are unjointed and do not possess claws (see Figs. 4.26 b and 4.26 c). Pseudopods are peculiar to larvae of butterflies and moths *(Lepidoptera)* and to larvae of sawflies *(Hymenoptera)*.

— Easily distinguishable spiracles (**not** included in Fig. 4.26 a).

Draw the spiracles on Fig. 4.26 a and label them.

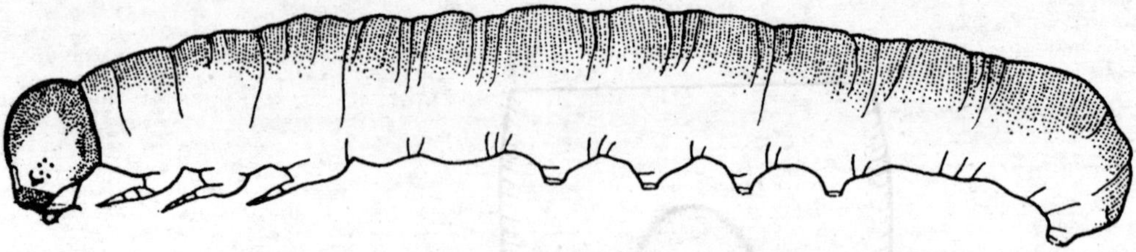

Label : Head, Thorax, Abdomen, True legs, False legs, Antenna, Mouth parts, Ocelli

Draw and label the spiracles
Length of insect to nearest mm. ________

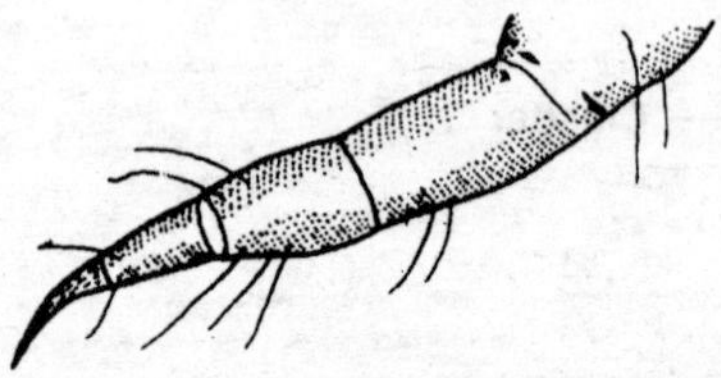

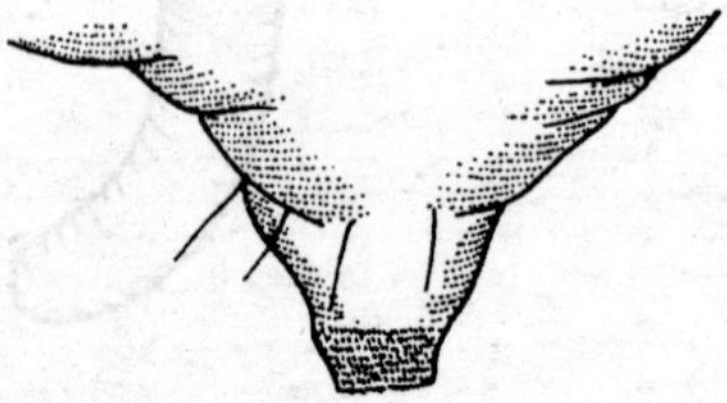

Fig. 4.26 (a) : Main external features of a Caterpillar (Larva of a moth), **(b)** Detail of true leg; **(c)** Detail of false leg

Answer the following questions with respect to moth larva:

1. How many ocelli are present on each side of the head? ________
2. On what segments of the abdomen are false legs present? ________
3. How false legs are different from true legs? ________
4. How many spiracles are there on:
   (*a*) the thorax ________
   (*b*) the abdomen ________
5. Are cerci present at the tip of abdomen? ________
6. How do the antennae differ from those of the cricket? ________

5

INTERNAL ANATOMY OF INSECTS

Insect internal anatomy consists of various organs performing different functions. These can be demonstrated by dissecting a common insect like cockroach, grasshopper or field cricket. Dissection should always be carried out under fluid, the insect completely submerged in it. Fresh specimens should be dissected in water while preserved specimens in 50 per cent alcohol. However, fresh insects may be dipped in 90 per cent alcohol before placing in water, because alcohol helps water to moisten the surface. Dissecting dish should be of dark base to make the dissection clearer. Short pins should be used for pinning the dissecting specimen. Since there is a nerve cord ventrally in all insects, they should be opened from **dorsal** side (so that nerve cord is not damaged).

Place the insect (e.g. cockroach) in a dissecting tray. Examine the details of its external anatomy and try to locate all of the features shown in the previous chapter. Carefully stretch out the wings and notice their appearance and texture. The tough, leathery forewings protect the more delicate, membranous hind wings. Before dissecting the body of the insect, cut off the wings at the base. Open the mouth using a probe and examine the mouth parts. With forceps and scalpel, separate the mouth parts and try to identify the labrum, mandibles, maxillae, and labium.

Now pin the insect with its dorsal side above in the dissecting dish. Pass one pin through the anterior region of the head and one pin through each of the metathoracic legs, and another pin through the posterior end of the abdomen. After pinning properly, cover the insect completely with water. Make a longitudinal cut mid-dorsally, from posterior end of the body to anterior end. Lift the cut edge of the body wall with fine forceps, being careful not to penetrate too deeply, and separate this from the underlying tissues with a scalpel. Pin down the body wall laterally on either side. Study the internal organs carefully and make a drawing of the general dissection (Fig. 5.1).

The tracheal system floats in the water as a mass of silvery, branched tubes. A female may have large yellow ovaries obscuring everything else; push these aside to find the dark-coloured stomach and the Malpighian tubules. Trace the complete digestive system. Along the ventral wall of the abdomen, you will find the double nerve cord with its enlarged ganglia, from which nerves branch to tissues.

The main internal organ systems of cockroach will be shown in a series of demonstrations.

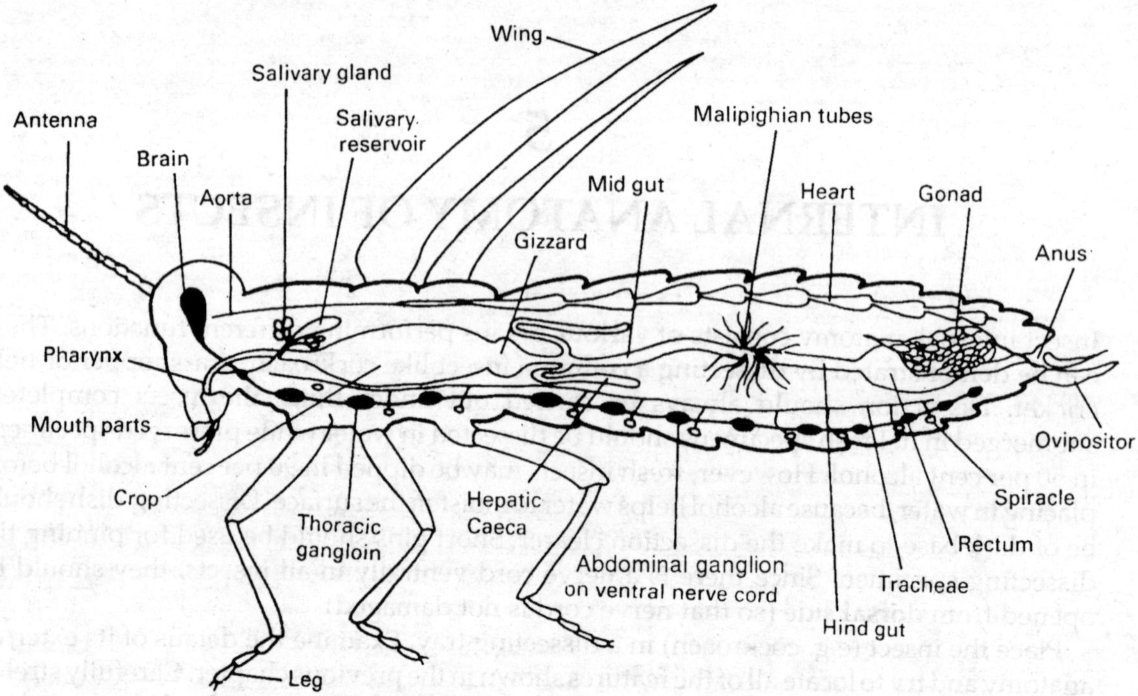

Fig. 5.1: Internal structure of an insect (diagrammatic)—lateral view.

Circulatory System

The entire internal body cavity (haemocoel) and contained organs, are bathed in body fluid (haemolymph). This body cavity may be divided longitudinally into sinuses by membranous diaphragms. There are few blood vessels. The main one is a longitudinal vessel close to the dorsal body wall of the insect (Fig. 5.2). This is muscular in the latter part and has a series of lateral inlet valves. This portion is known as the **heart** and it pumps the blood forward through the tubular forward extension which is without valves. This is referred to as **aorta** and is open at the head end. The heart is divided into chambers, which may be as many as 13 as in cockroaches, or only 3, as in the housefly. The pumping action of the heart can best be seen through the body wall of a live caterpillar.

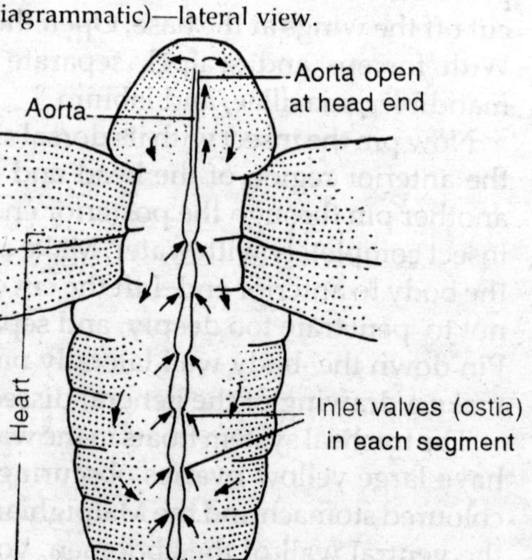

Fig. 5.2: Circulatory system of cockroach (dorsal view)

Respiratory System

Respiration in insects is direct from the atmosphere to the tissues through a series of tubes called **tracheae.** There are no organs comparable to lungs and the blood does not carry oxygen, and thus with few exceptions, is not pigmented. The openings to the exterior of the tracheal system are a series of holes along the sides of the insect referred to as **spiracles.** There are normally two main longitudinal trunks of tracheae from the spiracles. These also connect from side to side (Fig. 5.3). From these trunks, branched tracheae (tracheoles) go to all parts of the body, with finer and finer divisions so that all tissues are connected. Circulation of air through the tracheal system may be assisted by a series of air sacs and by movements of the insect's abdomen. The main tracheal trunks appear silver-white in a freshly-dissected or living specimen due to contained air.

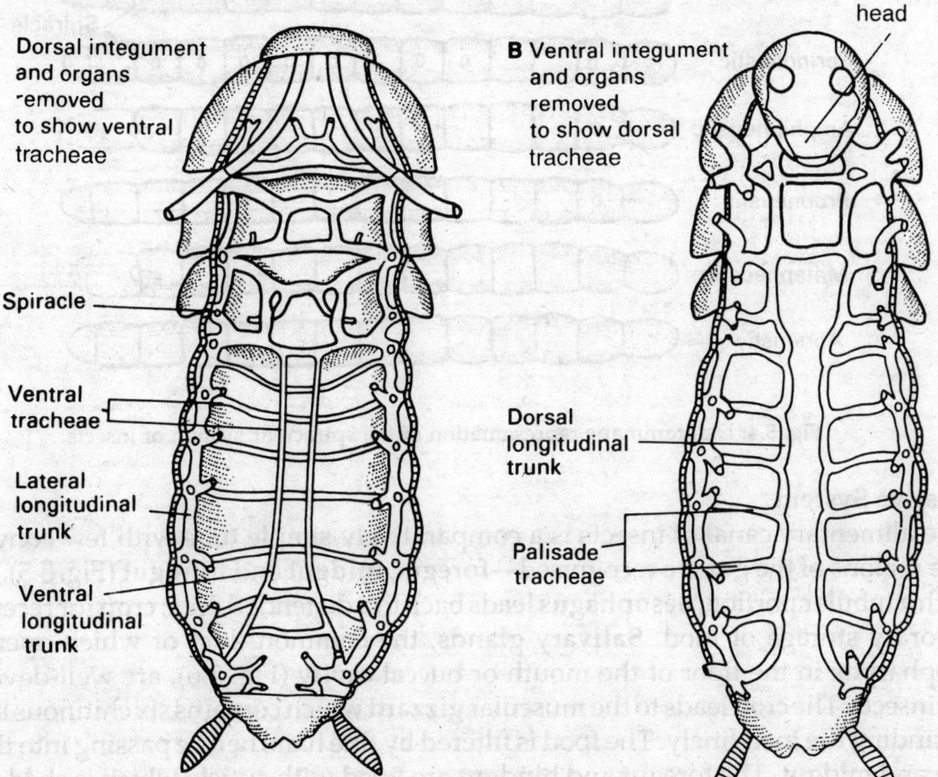

Fig. 5.3: Respiratoy system of cockroach (diagrammatic).

The distribution of spiracles varies in different insects. Most adult insects and larvae possess ten pairs but larvae sometimes have fewer spiracles (Fig. 5.4). Depending on the number and position of the functional, open spiracles, several types of tracheal system are characterized. In the *holopneustic* condition, there are all ten open pairs of spiracles, two on the thorax and eight on the abdomen. This is common in adult insects and many immature forms. The *peripneustic* condition has all but a few spiracles open while the *amphipneustic* type has only the first and last pairs open. The latter condition is found in the larvae of the higher Diptera. In *metapneustic* system the front pair is lost as seen in mosquito larvae and in *propneustic* system the last pair is lost (mosquito pupae). The *apneustic* type has no functional spiracles as in some aquatic insects where respiration occurs through the general body surface or gills. Some authors name those insects which have neither spiracles nor tracheae as *atracheate* type, e.g., springtails and some parasitic insect larvae.

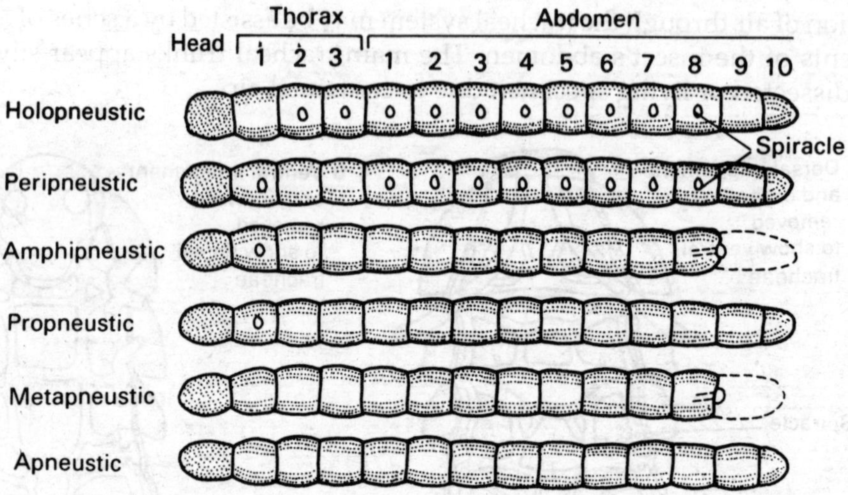

Fig. 5.4: Diagrammatic representation of the spiracular system of insects.

Digestive System

The alimentary canal of insects is a comparatively simple tube with few convolutions. Three regions of the gut are recognised—**foregut, midgut** and **hindgut** (Fig. 5.5). From the mouth a tubular portion, **oesophagus** leads back to a distended area, **crop** for reception and temporary storage of food. **Salivary glands**, the common duct of which opens on the hypopharynx in the floor of the mouth or buccal cavity (Fig. 5.6), are well-developed in most insects. The crop leads to the muscular **gizzard** which contains six chitinous teeth used for grinding the food finely. The food is filtered by fine hairs before passing into the enteric caeca and midgut. The foregut and hindgut are lined with cuticle which is shed when the insect moults. Digestion of food is completed in midgut and caeca and absorption of digested food can only take place in the midgut which is not lined with cuticle. Considerable absorption of moisture from digested food waste takes place in the hindgut and the waste material is stored in the rectum before it is expelled.

Excretory System

The excretory system consists of a series of thread-like tubes (*Malpighian tubes*) which collect nitrogenous wastes and discharge into the gut at about the junction of the mid-and hind-gut regions (Table 5.1). Identify the main regions of the gut and Malpighian tubes from the diagram (Fig. 5.1).

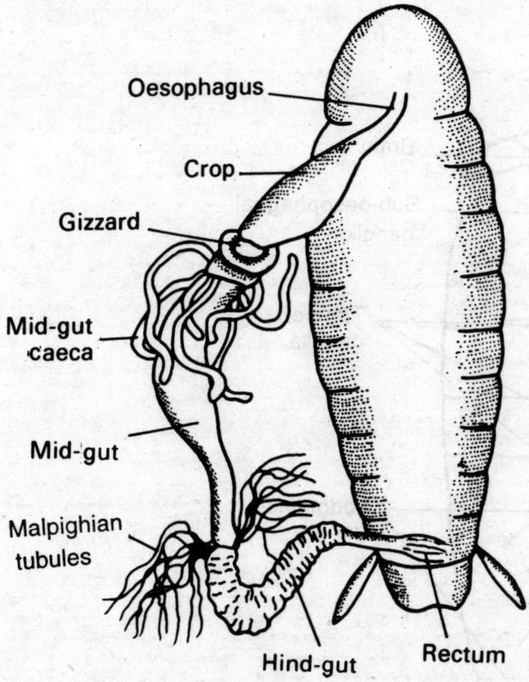

Fig. 5.5: Alimentary canal of cockroach (dorsal view).

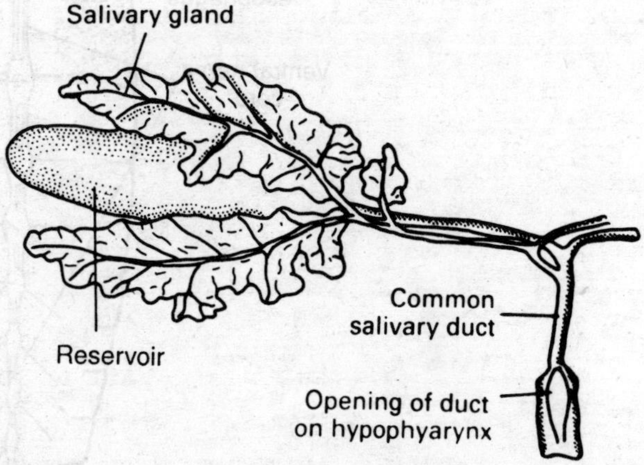

Fig. 5.6: Salivary gland from cockroach [*Periplaneta* (one side)].

Central Nervous System

This consists of a double nerve trunk running along the ventral side of the insect. Swellings in this nerve cord (ganglia) occur in each body segment. Branched nerves to various parts of the body arise from the ganglia. In the head region the nerve cord divides into two to run in a ring on either side of the oesophagus (Fig. 5.7). Three pairs of ganglia are fused into a mass of nervous tissue above the oesophagus to form the *brain* and three pairs fused below to form the *sub-oesophageal ganglion*. Identify the longitudinal ventral nerve cord in the dissected specimen and note the ganglia in each segment.

Reproductive System

The sexes are separate. In the male insect the internal organs consist of a pair of *testes* which lie in the abdomen (Fig. 5.8). Each testis is made up of small finger-like testis follicles. From each testis a duct (*vas deferens*) leads down to a storage reservoir for sperm (*seminal vesicle*). From each seminal vesicle a further duct connects to the male copulatory organ (*penis, aedeagus*) via a *central ejaculatory duct*. Paired *accessory glands* usually open into these latter ducts below the seminal vesicles.

In female insects the internal reproductive organs are paired *ovaries* that lie in the abdominal cavity (Fig. 5.9). Each ovary is made up of a number of similar units called *ovarioles*. There are three types of ovarioles (Fig. 5.10) :

(i) The panoistic type is the primitive type which has no nurse cells. The developing

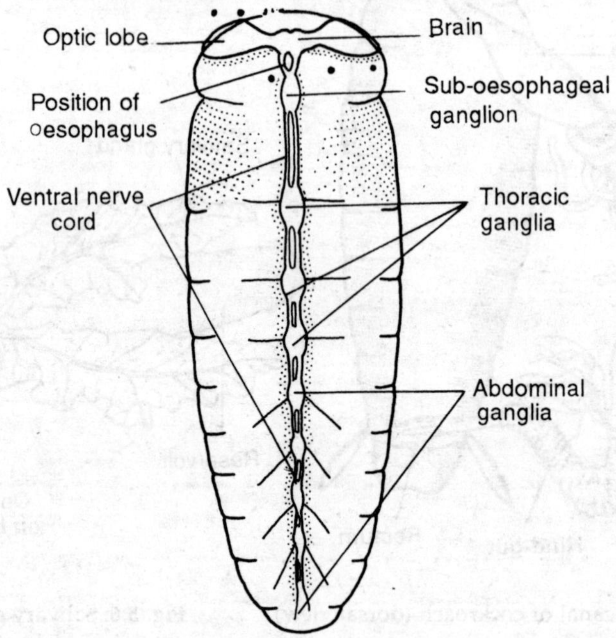

Fig. 5.7: Central nervous system of cockroach.

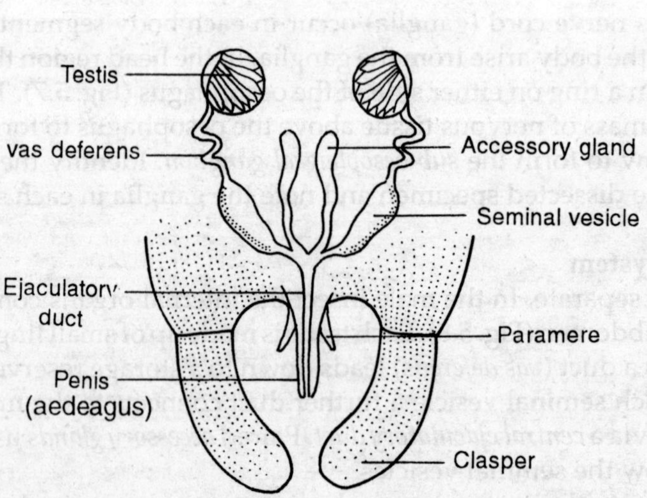

Fig. 5.8: Male reproductive system of an insect (diagrammatic) (after Snodgrass)

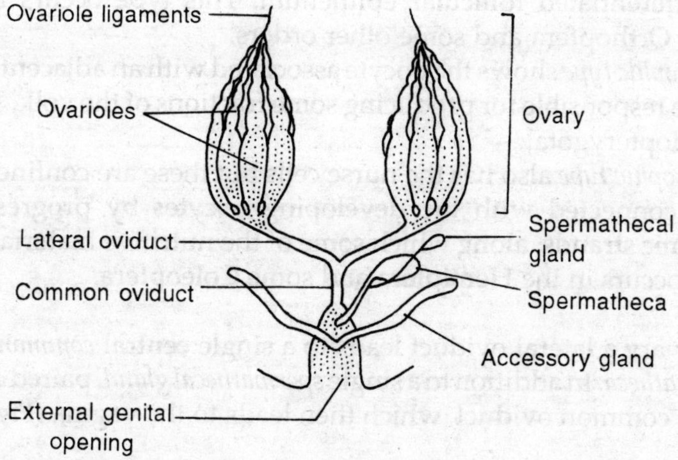

Fig. 5.9: Female reproductive system of an insect (diagrammatic) (after Snodgrass)

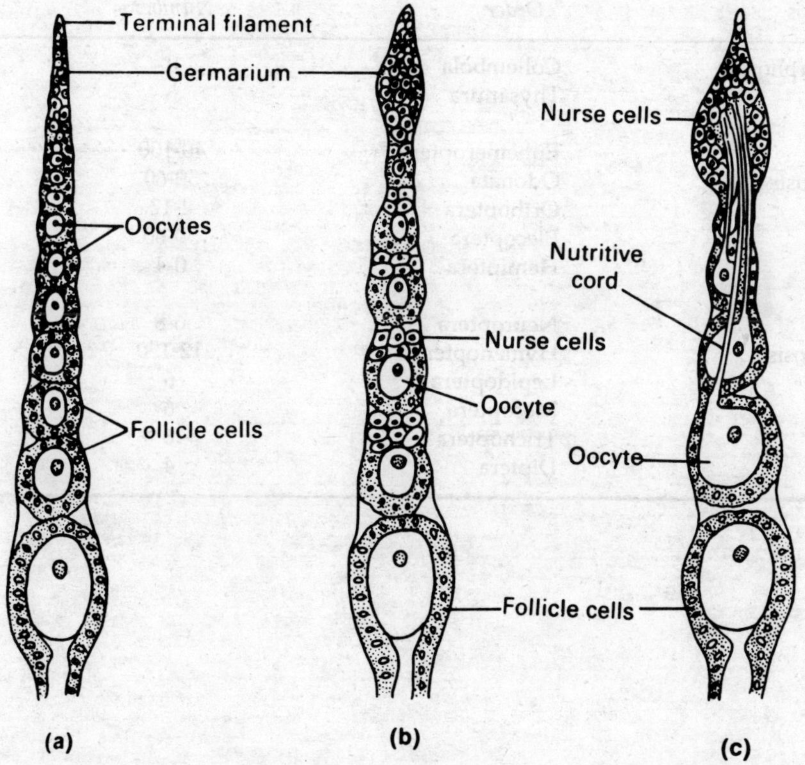

Fig. 5.10: Three main types of ovarioles, diagrammatic (after Webs)
(a) Panoistic type; (b) Polytrophic type; (c) acrotrophic type

oocyte may itself synthesize some components of the yolk and receives others from the undifferentiated follicular epithelium. This type occurs in the Apterygotes, Odonata, Orthoptera and some other orders.

(ii) *The polytrophic type* shows the oocyte associated with an adjacent group of nurse cells which are responsible for producing some fractions of the yolk. This type is found in most Endopterygota.

(iii) *The acrotrophic type* also has the nurse cells but these are confined to the germarium and are connected with the developing oocytes by progressively lengthening cytoplasmic strands, along which some of the nutritive materials pass. This kind of ovariole occurs in the Hemiptera and some Coleoptera.

From each ovary a lateral oviduct leads to a single central *common oviduct*, into which opens the *spermatheca*. In addition to a single *spermathecal gland*, paired *accessory glands* open into the central common oviduct, which then leads to the vagina (Fig. 5.9).

TABLE 5.1. Representative Numbers of Malpighian Tubules as Related to Metamorphosis

Type of Metamorphosis	*Order*	*Number*
No Metamorphosis	Collembola	0
	Thysanura	6
Incomplete Metamorphosis	Ephemeroptera	40-100
	Odonata	50-60
	Orthoptera	2-12
	Plecoptera	50-60
	Hemiptera	0-4
Complete Metamorphosis	Neuroptera	6-8
	Hymenoptera	12-150
	Lepidoptera	6
	Mecoptera	6
	Trichoptera	6
	Diptera	4

6

DEVELOPMENT AND METAMORPHOSIS

TYPES OF METAMORPHOSIS

The most striking feature of insect development is the marked change of bodily form which most insects undergo from young to adult, a process known as *metamorphosis*. All insects can be grouped into three categories, according to their type of development.

1. Ametabola (Apterygota)

There is no appreciable change in form with growth though smaller stages are not sexually mature (Fig. 6.1 a). Moulting may continue once adult. Wings are never present.

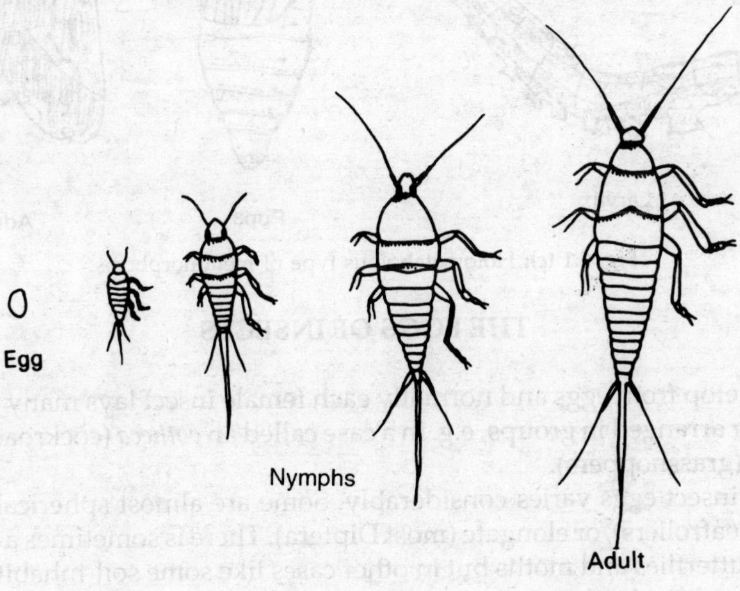

Fig. 6.1 (a): Ametabolous type of metanophosis.

2. Hemimetabola (Exopterygota)

Metamorphosis is incomplete, i.e., young stages (nymphs) are similar in form to adults, possessing similar mouth parts and feeding in the same fashion (Fig. 6.1 b). Nymphs may possess compound eyes. There is no pupal stage. Wings develop externally.

3. Holometabola (Endopterygota)

Metamorphosis is complete with both larval and pupal stages. Young stages (larvae) are very different in form from adults, possessing different mouth parts and feeding in a different fashion (Fig. 6.1 c). Larvae never possess compound eyes. Wings develop internally in pupa.

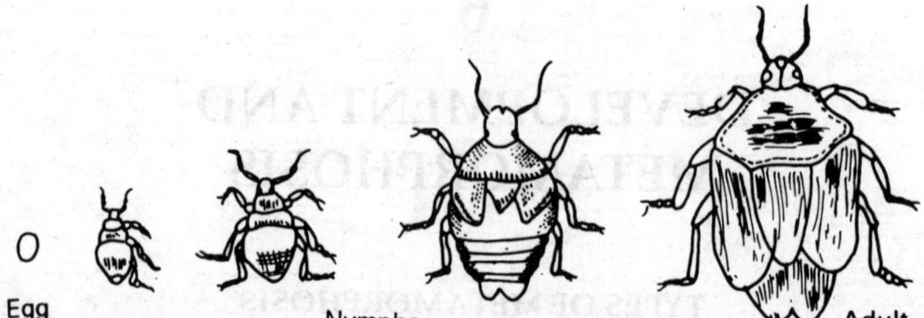

Fig. 6.1 (b): Hemimetabolous type of metamorphosis.

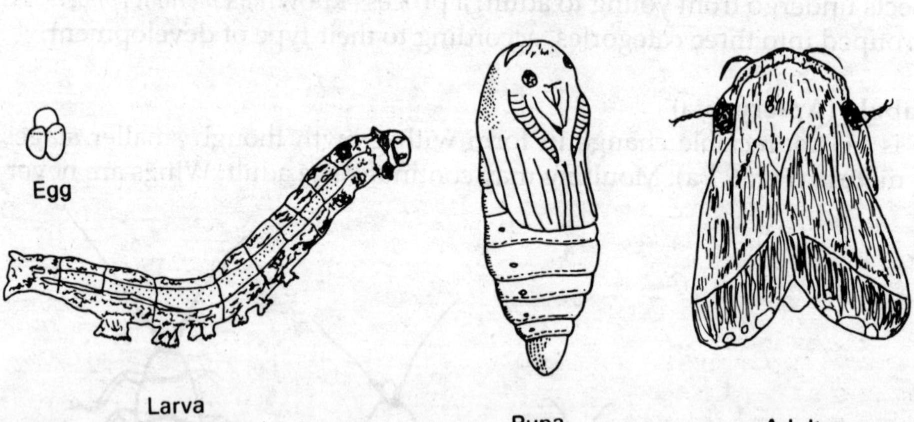

Fig. 6.1 (c): Holometabolous type of metamorphosis.

THE EGGS OF INSECTS

All insects develop from eggs and normally each female insect lays many eggs. Eggs may be laid *singly* or arranged in groups, e.g. in a case called an *ootheca* (cockroaches), or in a so-called *egg-pod* (grasshoppers).

The *form* of insect eggs varies considerably. Some are almost spherical but others are flattened (e.g. leafrollers), or elongate (most Diptera). There is sometimes a definite shell to the egg as in butterflies and moths but in other cases like some soil-inhabiting beetles, the eggs are quite soft and must absorb moisture before hatching. Collect insect eggs from leaves and stems of plants or from the soil, and make their drawings to show the range of variation in structure, form and colouration. Also record the place in which it is found; the way in which it is laid or attached to substratum. This sometimes may be of great importance in forecasting the appearance of the destructive stages of a particular insect pest.

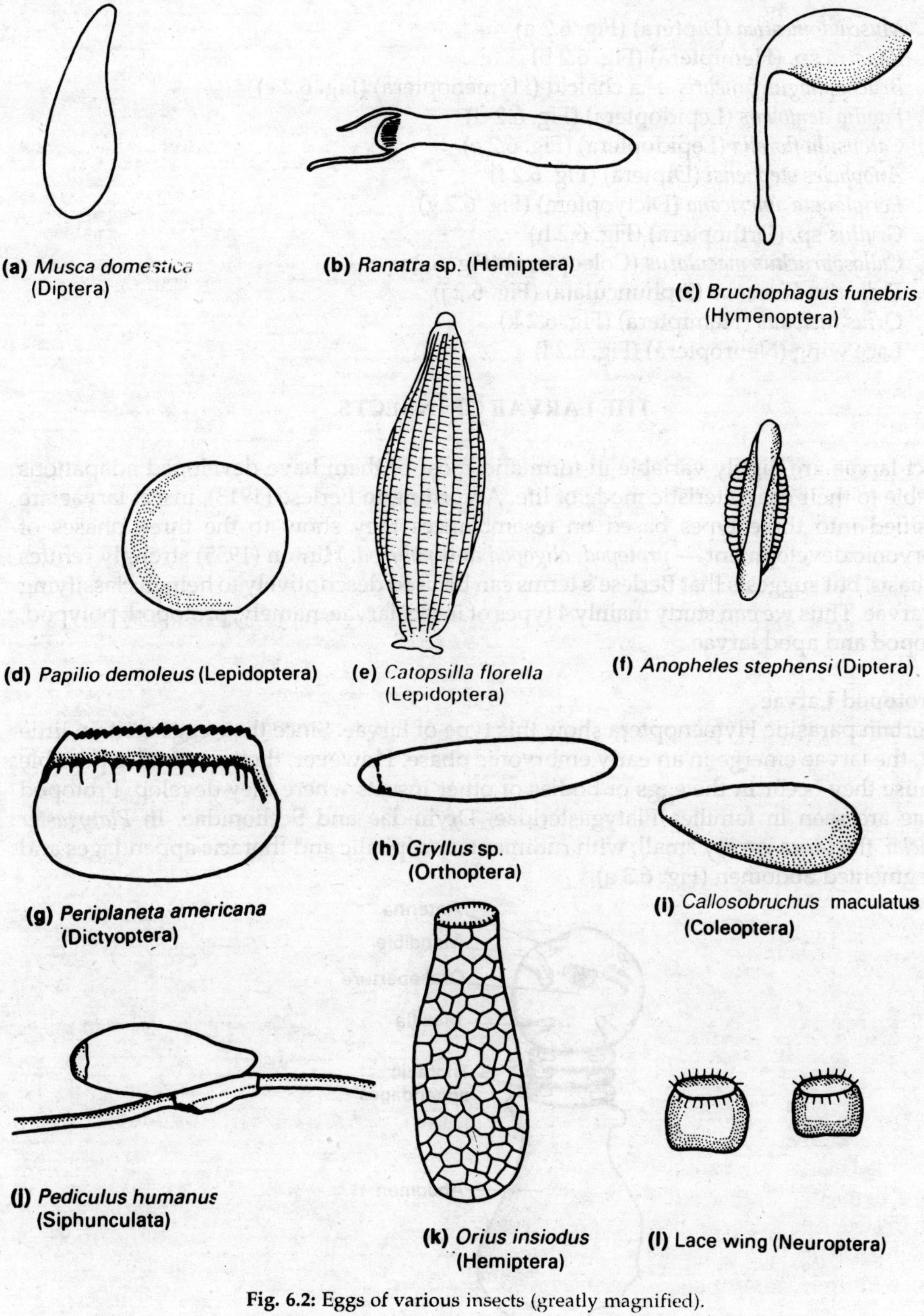

Fig. 6.2: Eggs of various insects (greatly magnified).

1. *Musca domestica* (Diptera) (Fig. 6.2 a)
2. *Ranatra* sp. (Hemiptera) (Fig. 6.2 b)
3. *Bruchophagus funebris* — a chalcid (Hymenoptera) (Fig. 6.2 c)
4. *Papilio demoleus* (Lepidoptera) (Fig. 6.2 d)
5. *Catopsilla florella* (Lepidoptera) (Fig. 6.2 e)
6. *Anopheles stephensi* (Diptera) (Fig. 6.2 f)
7. *Periplaneta americana* (Dictyoptera) (Fig. 6.2 g)
8. *Gryllus* sp. (Orthoptera) (Fig. 6.2 h)
9. *Callosobruchus maculatus* (Coleoptera) (Fig. 6.2 i)
10. *Pediculus humanus* (Siphunculata) (Fig. 6.2 j)
11. *Orius insiodus* (Hemiptera) (Fig. 6.2 k)
12. Lace wing (Neuroptera) (Fig. 6.2 l)

THE LARVAE OF INSECTS

Insect larvae are highly variable in form and most of them have developed adapations suitable to their characteristic mode of life. According to Berlese (1913), insect larvae are classified into three types based on resemblances they show to the three phases of embryonic development — *protopod, oligopod* and *polypod*. Hinton (1955) strongly refutes this basis, but suggests that Berlese's terms can be used descriptively to help in classifying the larvae. Thus we can study mainly 4 types of insect larvae, namely, protopod, polypod, oligopod and apod larvae.

1. Protopod Larvae

Certain parasitic Hymenoptera show this type of larvae. Since their eggs contain little yolk, the larvae emerge in an early embryonic phase. However, their survival is possible because they occur in the eggs or bodies of other insects where they develop. Protopod larvae are seen in families Platygasteridae, Dryinidae and Scelionidae. In *Platygaster herrickii,* the larva is very small, with rudimentary cephalic and thoracic appendages and unsegmented abdomen (Fig. 6.3 a).

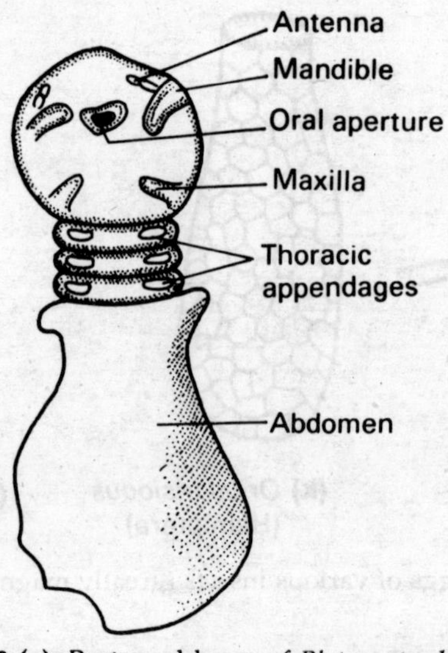

Fig. 6.3 (a): Protopod larva of *Platygaster herrickii.*

2. Polypod Larvae

Polypod or *eruciform* larvae (Fig. 6.3 b) are seen in Lepidoptera and Mecoptera. Their body is cylindrical, clearly segmented with occurrence of abdominal limbs or prolegs. The antennae and thoracic legs are present, though not well-developed. They are usually inactive and live in the vicinity of their food.

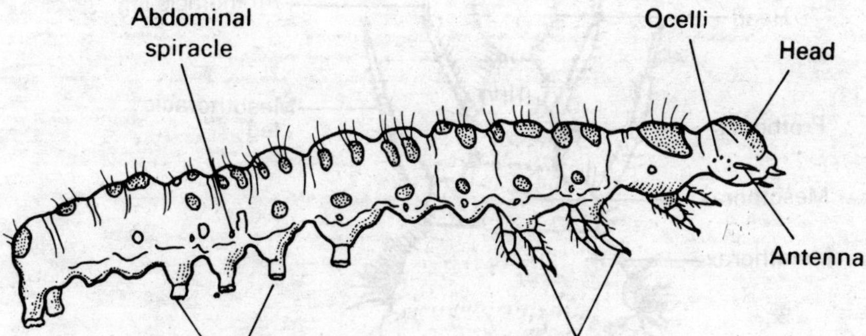

Fig. 6.3 (b): Polypod or eruciform larva of *Pieris brassicae*

3. Oligopod Larvae

These are characterized by the absence of abdominal prolegs though a pair of caudal processes or urogmphi may be present. Thoracic legs are well-developed. Oligopod larvae may be distinguished into these three types :

(a) *Campodeiform larvae*: These have long, dorso-ventrally flattened, well-sclerotized body with prognathus head. They usually have long thoracic legs and a pair of terminal abdominal processes. They are very active predators with well developed sensory apparatus. Such larvae occur in Neuroptera (Fig.6. 3 c), Trichoptera, Strepsiptera and some Coleoptera (Adephaga).

(b) *Carabiform larvae*: These are similar to campodeiform type, but the legs are shorter and filaments are lacking on the end of abdomen. Examples are seen in carabid and chrysomelid beetles (Coleoptera).

(c) *Scarabaeiform larvae*: These have C-shaped, stout sub-cylindrical and soft body. They have a well-developed head, usually with short thoracic legs but no prolegs. They generally lead a sluggish life living in the ground. They are characteristic of Scarabaeoidea (Coleoptera) (Fig. 6.3 d).

4. Apod Larvae

In these larvae the trunk appendages are completely absent, with poorly developed head and without eyes. There are three types of apodous larvae based on the degree of sclerotization of the head:

(a) *Eucephalous:* They have more or less well sclerotized head-capsule with relatively little reduction of the cephalic appendages, e.g., Nematocera (Diptera) (Fig. 6.3 e).

(b) *Hemicephalous:* The head-capsule and its appendages are reduced that can be retracted into the thorax, e.g., Brachycera (Diptera).

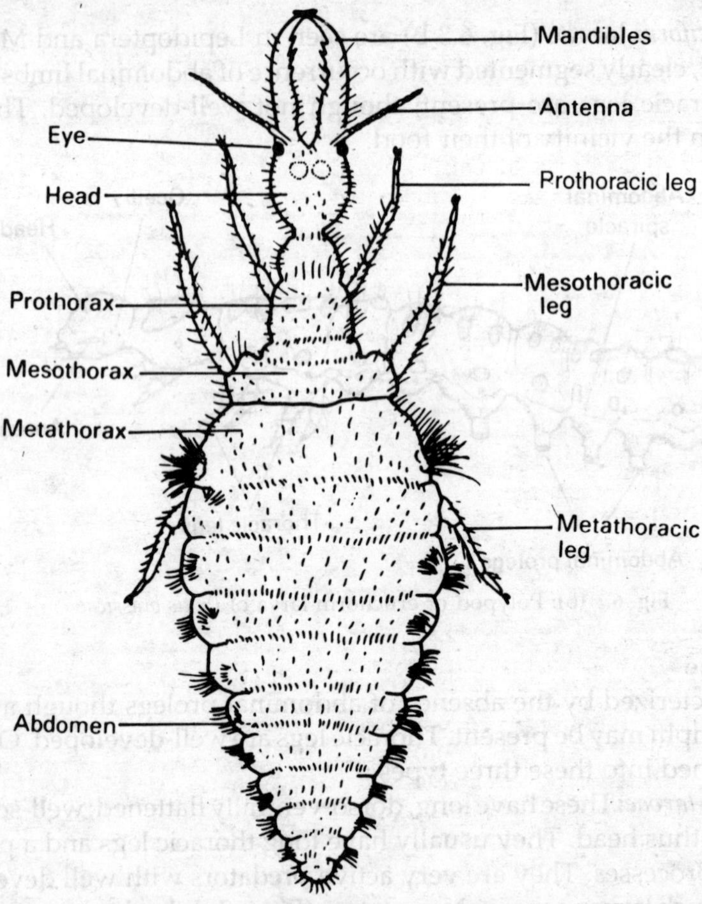

Fig. 6.3 (c): Oligopod-campodeiform larva of *Hagenomiya tristic* (Neuroplera)

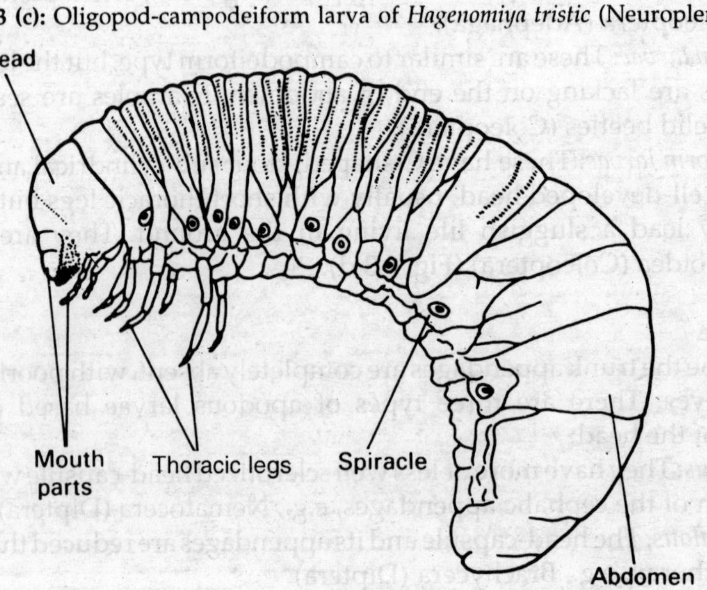

Fig. 6.3 (d): Oligopod-scarabaeiform larva of scarabaeid beetle.

(c) *Acephalous:* There is no head capsule as such, though some cephalic structures may be present, e.g., Cyclorrhapha (Diptera) (Fig. 6.3 f).

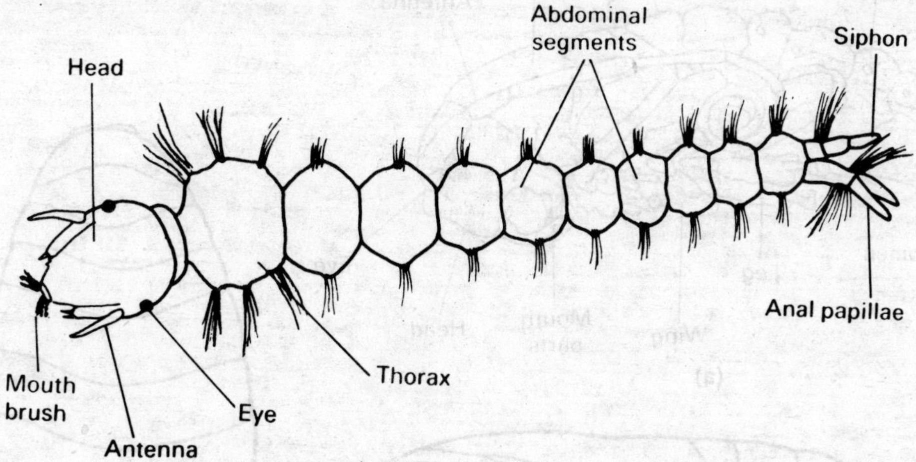

Fig. 6.3 (e): Apodous-eucephalus larva of *Culex fatigans.*

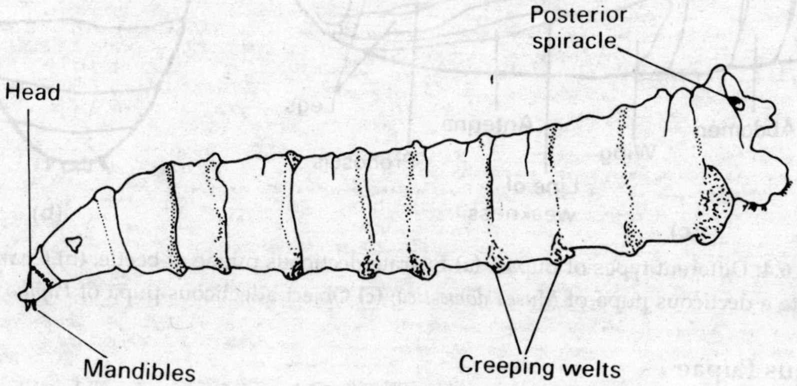

Fig. 6.3 (f): Apodous-acephalous larva of *Musca domestica.*

The Pupae of Insects

Based on the adherence of their appendages to the body, insect pupae are classified into two main types by Hinton (1946).

1. Decticous Pupae

These are of the primitive type and are always *exarate*, i.e., the appendages are not adherent to the rest of the body and can be used in locomotion. They have powerful, sclerotized, articulated mandibles which are used to escape from the pupal cell. Examples of exarate type of pupa are seen in Neuroptera, Mecoptera, most Trichoptera, Lepidoptera and Coleoptera (Fig. 6.4 a).

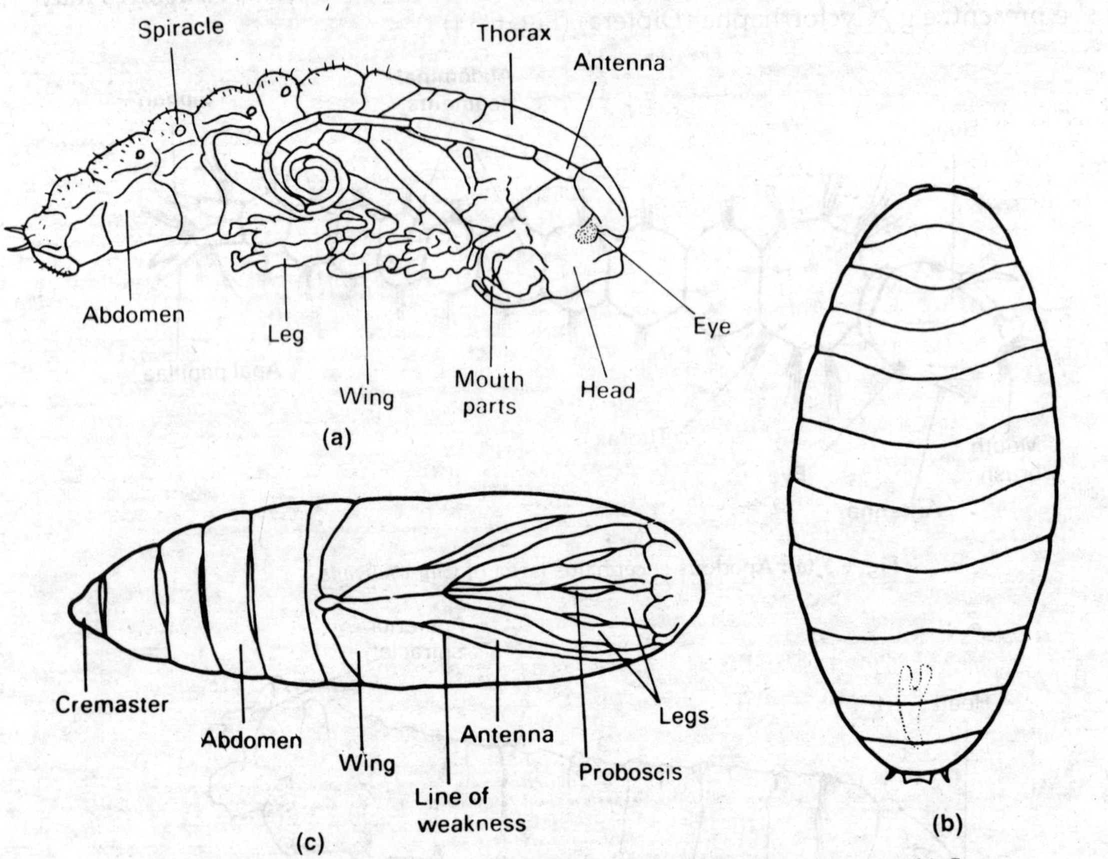

Fig. 6.4: Different types of pupae. **(a)** Exarate decticous pupae of beetle; **(b)** Coarctate or exarate a decticous pupa of *Musca domestica*; **(c)** Object adecticous pupa of *Papilio domoleus*.

2. Adecticous Pupae

Pupae of this type have non-articulated mandibles, often reduced and are not used in escaping from the pupal cell. Two main types of adecticous pupae are :

(a) Exarate adecticous pupae: Also known as coarctate pupae in Cyclorrhaphan Diptera (Fig. 6.4 b) where pupa is enclosed in a puparium which is formed from a preceding larval cuticle. In these pupae also, appendages are free of any secondary attachment to the body, e.g., Cyclorrhapha (Diptera), Siphonaptera, Strepsiptera, Coleoptera and Hymenoptera.

(b) Obtect adecticous pupae: The appendages of the pupa are close to the body, held there by a tight-fitting outer envelope. The exposed surfaces of the appendages are more heavily sclerotized. Examples are seen in Lepidoptera (Fig. 6.4 c), Nematoceran and Brachyceran Diptera, Coleoptera and many Hymenoptera.

7

INSECT CLASSIFICATION

About thirty insect orders are recognized but the exact number will depend on which authority is followed. In this book the classification used is that of Imms (10th edition, 1977). An outline of the classification of insects is set out in Table 7.1. The grouping of orders into sub-classes is based on the type of metamorphosis (if any), as discussed in Chapter 6.

Insects are important in agriculture and horticulture as pests of plants (or as pests of livestock), as natural enemies of pests, and in a few cases as pollinators. Those insects of importance in relation to cultivated plants come from 9 or 10 orders (Table 7.1). Insects from 5 or 6 orders are represented by a few rare species and of the remainder, several are of no pest significance or are exclusively pests of animals rather than of plants (e.g. lice, fleas).

What you are Expected to Know

You should know the scientific and common names of those insect orders which are of practical importance in relation to plants or animals. You should also be able to place each order into its correct sub-class within the insects, i.e., ametabola, hemimetabola or holometabola.

TABLE 7.1: An Outline Classification of Insects

Subclass I AMETABOLA (= APTERYGOTA)

Adults wingless (primitive condition). No metamorphosis. Moulting continues once adult.

	Common names
Order 1. Thysanura	— three-pronged bristle tails, silverfish
2. Diplura	— two-pronged bristle tails
3. Protura	— proturans
4. Collembola	*— springtails

Subclass II METABOLA (= PTERYGOTA)

Adults usually winged (but may be secondarily wingless). Metamorphosis occurs. Do not moult once adult (except for mayflies).

Division A. HEMIMETABOLA (= EXOPTERYGOTA)

Metamorphosis incomplete. No true pupal stage. Wings develop externally in nymphal stages.

Order		
	5. Ephemeroptera	—mayflies
	6. Odonata	—dragonflies
	7. Plecoptera	—stoneflies
	8. Grylloblattoidea	— —

	Order		
	9. Orthoptera	**	—crickets, grasshoppers, locusts
	10. Phasmida	*	—stick insects, leaf insects
	11. Dermaptera	*	—earwigs
	12. Embioptera	—	—
	13. Dictyoptera	×	—cockroaches, mantids
	14. Isoptera	×	—termites
	15. Zoraptera	—	—
	16. Psocoptera		—psocids, booklice
Phthiraptera	17. Mallophaga	+	—biting lice, bird lice
Phthiraptera	18. Siphunculata	+	—sucking lice
	19. Hemiptera	**	—cicadas, leafhoppers, scale insects, aphids, plant bugs, etc.
	20. Thysanoptera	*	—thrips

Division B. HOLOMETABOLA (=ENDOPTERYGOTA)

Metamorphosis complete. Pupal stage present. Wings develop internally in pupal stage.

	Order		
Order	21. Neuroptera	*	—lacewings, ant lions
	22. Coleoptera	**×	—beetles
	23. Strepsiptera		—stylopids
	24. Mecoptera		—scorpion flies
	25. Siphonaptera	+	—fleas
	26. Diptera	**+×	—flies
	27. Lepidoptera	**×	—butterflies, moths
	28. Trichoptera		—caddis flies
	29. Hymenoptera	**	—sawflies, ants, bees, wasps

For the 16 or 17 orders you should have an understanding of their main characteristics and why each group is primarily of importance (as pests of plants, as natural enemies of such pests, as pollinators, or as animal pests). Within each order you should be familiar with one or two important representative species so that you could:

(i) recognise the main life stages
(ii) describe the mode of feeding, nature of plant injury and plants attacked in case of plant pests and nature of harm done and diseases transmitted in case of animal pests.
(iii) understand the life cycle, mode of reproduction and generations per year.
(iv) discuss both natural controlling factors and artificial control measures.

Finally, you should aim to be on at least nodding acquaintance with all the main pest species in each order — ability to recognise the main life stages; knowledge of plants attacked; damage symptoms; and the diseases transmitted.

Practical Classes in Insect Orders

Each practical class will normally deal with one or two orders depending on their

** — orders of major economic importance in relation to cultivated plants.
* — orders of limited economic importance in relation to cultivated plants.
+ — orders of importance as parasites of vertebrate animals.
× — orders important as pests of timber, of stored foods, or of public health significance.

importance. In each case the material for study will be divided into:

(*a*) material for personal study at your allotted bench place (bench work), and

(*b*) display material which will be located on one of the outer benches of the laboratory.

Material for personal study will be provided as individual labelled specimens at your bench place or at clearly marked positions around the laboratory. Examine this material *non-destructively* as far as possible and return specimens to their containers when you have finished with them. However, if material is unavoidably damaged, do not return it, but dispose of it in the waste containers provided.

How to Proceed

Benchwork

1. Read the notes appropriate to the practical class and study any drawings or diagrams provided.
2. Examine the specimens (with the unaided eye and/or with a microscope) and identify all the features listed.
3. Complete the drawings and labelling as indicated.

Display

1. Study the display material (specimens, photographs and literature), paying particular attention to recognition of the main life stages of each pest and to symptoms of plant injury, or animal diseases.
2. Collect copies of information sheets on individual pests. Answer the questions listed.

You will study the following orders in this exercise:

1. Order Collembola
2. Order Thysanura
3. Order Orthoptera
4. Order Phasmida
5. Order Dermaptera
6. Order Dictyoptera
7. Order Isoptera
8. Order Mallophaga
9. Order Siphunculata
10. Order Hemiptera
11. Order Thysanoptera
12. Order Neuroptera
13. Order Lepidoptera
14. Order Coleoptera
15. Order Hymenoptera
16. Order Diptera
17. Order Siphonaptera

ORDER COLLEMBOLA

Common name: springtails

Examine specimens of **springtails** under the microscope and observe the following characteristic features of all Collembola:

- absence of wings
- little or no metamorphosis
- mouthparts (of biting type) inserted into the head and difficult to see
- no compound eyes
- antennae normally four or five jointed

- abdomen six segmented in elongate forms, without obvious segmentation in globular forms
- spring on ventral surface of abdomen (not present in all species)
- ventral tube on ventral surface of first abdominal segment (involved in uptake of water)
- no spiracles
- legs terminate in simple claws without jointed tarsi.

Most species of springtails require high humidity and occur in the soil and other moist conditions. Since they have no tracheal system they respire through the general body surface. Generally only young, tender plant parts are attacked, mostly at the seedling stage. Small holes are eaten in leaves, cotyledons or shoots.

There are two groups of springtails:

- elongate (Arthropleona)
- globular (Symphypleona)

In the latter, the abdomen is almost spherical in shape with the segments apparently merged together. Most pest species are of globular type, e.g., *Sminthurus viridis* which is a pest of pasture in tropics. They feed on the mesophyll layer between the upper and lower epidermis of leaves, and thus the leaves become transparent.

Provide headings for the Figs. 7.1 and 7.2 and indicate the approximate size (in mm) of the actual insects. Indicate the colour of each. (Insects that have been preserved tend to lose their colour, so record the colour of the insect *when alive.* Springtails are generally pale greenish yellow).

Examine the display material and answer the following questions:

1. Is specimen A a springtail? Give reasons for your answer. ______________

____

2. To what subclass of insects do springtails belong? ______________

____

3. Are all species of springtails pests of plants? ______________

____

4. What is the nature of the plant injury caused by *Sminthurus viridis*?

____

5. How do the juvenile stages of springtails differ from the adults?

____

6. What features would you look for in order to recognise springtails *in the field*? ___

____

7. Name two features possessed by springtails and not by other insects.

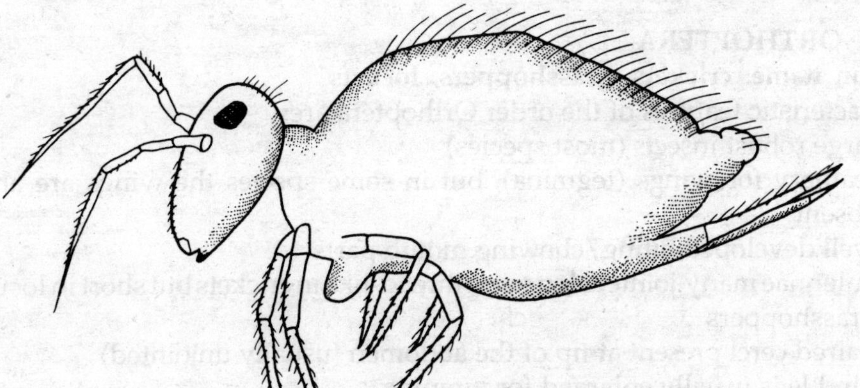

Size : ____________
Colour : ____________

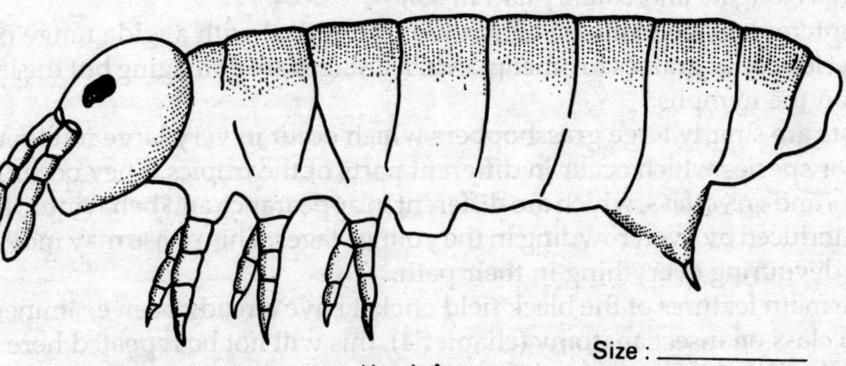

Size : ____________
Colour : ____________

***Label on each drawing* : Head, Antenna, Ocelli, legs, Spring, Ventral tube**

Fig. 7.1 & 7.2: Measure the size, fill the colour and label the drawings: head, antennae, ocelli, legs, spring, ventral tube.

ORDER THYSANURA

Common name: silverfish, three-pronged bristle tails

Examine a specimen of a silverfish and observe the following features:

- wings absent
- antennae long, thread-like with many joints
- mouth parts biting / chewing type
- compound eyes absent or present in a reduced form
- abdomen 11-segmented and bearing pairs of small appendages on some segments
- three long tail filaments, which are jointed.

Collect specimens of silverfish (*Lepisma* sp.) from old books and make a permanent preparation of it. Observe under a compound microscope and make a drawing of it in your notebook.

Answer the following question:

1. How will you differentiate between a three-pronged bristle tail and a two-pronged bristle tail? ____
____

ORDER ORTHOPTERA

Common name: crickets, grasshoppers, locusts

Characteristic features of the order Orthoptera are:

- large robust insects (most species)
- leathery forewings (tegmina), but in some species the wings are abbreviated or absent
- well developed biting/chewing mouth parts
- antennae many-jointed, long and thread-like in crickets but short in locusts and some grasshoppers
- paired cerci present at tip of the abdomen (usually unjointed)
- hind legs usually enlarged for jumping
- a well developed ovipositor present in the female but not in some grasshoppers
- eggs elongate and usually laid in soil

Orthoptera damage plants of all kinds and can deal with a wide range of plant tissue. Most species prefer grasses to other plants. All stages are damaging but the large adults eat more than the nymphs.

Locusts are simply large grasshoppers which occur in very large numbers. There are 6 or 7 major species which occur in different parts of the tropics. They occur in two phases — *solitary* and *gregarious*, which are different in appearance and behaviour. The gregarious phase is induced by overcrowding in the young stages. This phase may move in enormous swarms devouring everything in their path.

As the main features of the black field cricket have already been examined in an earlier practical class on insect anatomy (chapter 4), this will not be repeated here.

Answer the following questions:

1. Does specimen provided belong to the Orthoptera? Give reasons for your answer. ___

___

2. How many generations a year does black field cricket have? ___

___

3. In what life stage does black field cricket pass the winter? ___

___

4. At what time of year is plant damage most severe from black field cricket? ___

___

ORDER PHASMIDA

Common name: stick insects, leaf insects

Phasmids generally look like dead sticks or leaves of plants and this mimicry is advantageous because of its protective value. They are not very common. An adult phasmid will show the following characteristics:

- large insects of elongate cylindrical form (stick insects) or flattened (leaf insects).
- winged or wingless. Forewings when present usually small.

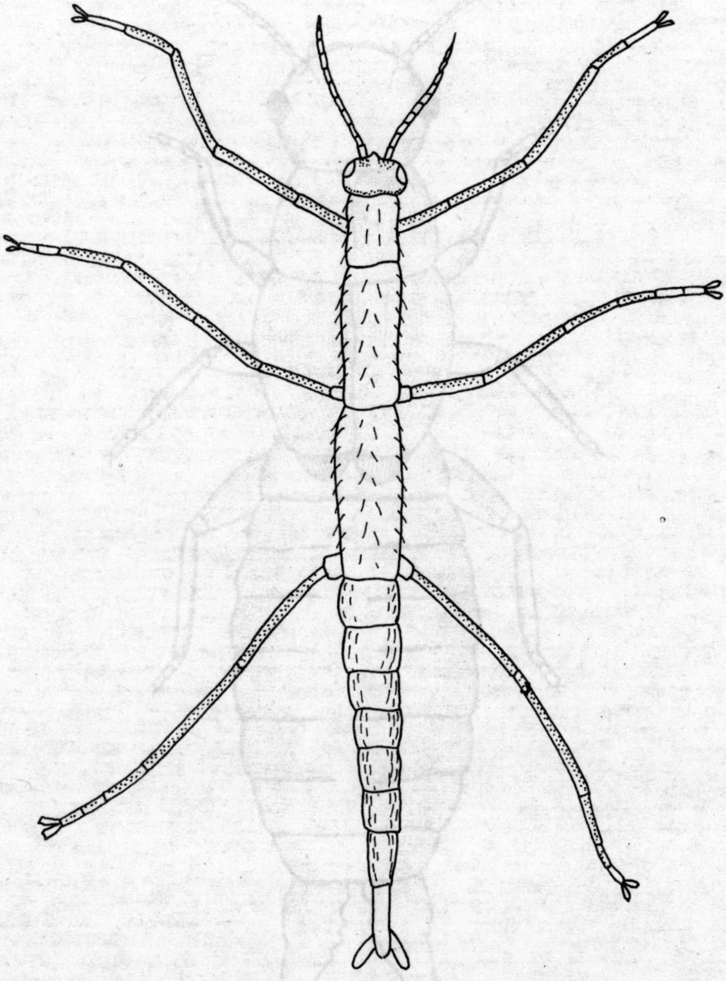

Size : ____________________

Colour : ____________________

Label the drawing

Fig. 7.3

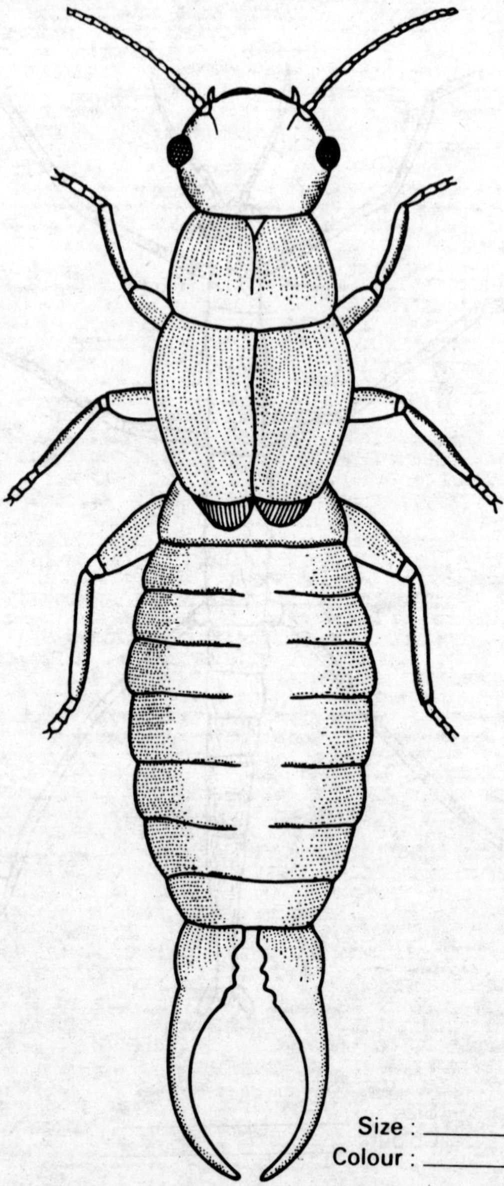

Size : ____________

Colour : ____________

Label the drawing : Tail, Forceps, Wing cases, Tips of hind wings Antenna, Thorax, Abdomen

Fig. 7.4

- middle and hind segments of thorax in stick insects very long.
- legs similar to each other, with five segmented tarsi.
- a pair of anal cerci, short and unjointed.
- mouth parts biting/chewing type.
- antennae filiform or moniliform; short or very long.
- ovipositor small, concealed by eighth abdominal sternum.
- all are plant feeders and some species of stick insects are plant pests in tropics.

Examine the Fig. 7.3 and provide a heading to it. Label the diagram and write its size.

Answer the following questions:

1. Why phasmids are called as stick insects? ______________________________

____

2. Which part of the plant they feed on? ______________________________

____

ORDER DERMAPTERA

Common name: earwigs

Following features characterize the order Dermaptera:

- mouth parts of basic biting/chewing pattern
- forewings reduced to small leathery flaps (tegmina)
- hindwings are almost semi-circular in shape and normally folded in a complex manner beneath the tegmina. They have an almost radial venation when spread out.
- cerci in the form of horny unjointed forceps.

Earwigs are relatively minor pests causing damage to the flowers of such plants as chrysanthemums and the fruits of soft and stone fruits. Damage may be caused by both the adult and immature stages. In some situations earwigs are predatory on other plant pests and thus regarded as beneficial. Earwigs are much more important in horticultural than in agricultural crops.

Provide a suitable heading to Fig. 7.4 and label it. Indicate size and colour.

Examine the prepared slide mounts of:

(a) hind wing of an earwig to show 'radial' pattern of veins, and

(b) tail forceps of male and female. Note that there is a difference in shape between the sexes.

Complete the following:

1. Why are earwigs sometimes regarded as beneficial? ______________________________

____

2. Can earwigs fly? Give reasons for your answer. ______________________________

____

ORDER DICTYOPTERA

Common name: cockroaches, mantids

Cockroach can be studied as a representative of this order. Chief features of the order Dictyoptera are:

- medium-sized to large insects with well developed exoskeleton.
- antennae long, filiform.

- mouth parts of generalised biting pattern.
- usually with two pairs of wings; forewings rather heavier in texture than hindwings; some species wingless.
- legs with large coxae and similar to each other in cockroach; forelegs modified for clasping in the mantids.
- females have concealed, reduced ovipositor; eggs produced in an ootheca.

Provide a heading to the diagram in the Fig. 7.5 and label the various parts.

Examine and draw the ootheca and the immature stages of the cockroach. Also remove one raptorial fore leg from the preying mantis and draw in detail, noting the spines present on the femur and tibia.

The cockroaches and mantids belong to the suborders *Blattaria* and *Mantodea*, respectively.

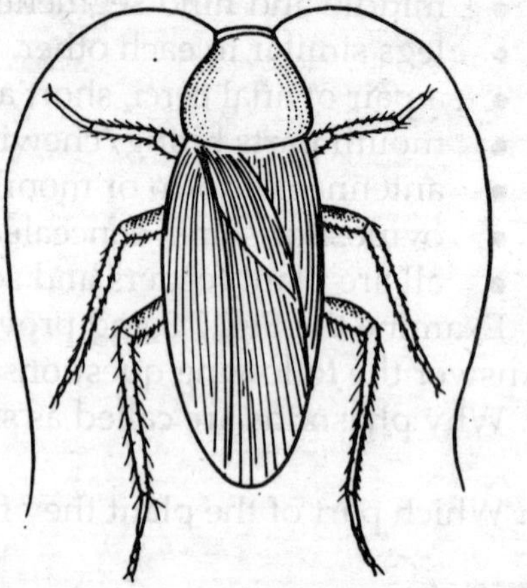

Size : ____________

Colour : ____________

Fig. 7.5:

Cockroaches are nocturnal, omnivorous scavengers consuming human food of all kinds as well as decaying organic matter. Some species occasionally attack growing plants. The mantids on the other hand are entirely predatory on insects and other small arthropods. They are therefore regarded as beneficial. Even the nymphs which hatch out of the eggs are predacious. Most species are cannibalistic and it is known that the females eat the male after copulation.

Answer the following questions:

1. How would you distinguish a cockroach from a mantid? ____________
2. Why mantids are considered beneficial to man? ____________
3. Where would you look for a mantid? ____________

ORDER ISOPTERA

Common name: termites

All termite species live in complex colonies throughout the tropics and in warm temperate countries. Individuals of the colony are morphologically adapted to their different functions in the colony. The castes are composed of a few winged reproductive forms and large numbers of wingless sterile workers and soldiers.

Examine the different morphs of termites and make labelled drawings of the following:

(*a*) Worker

(*b*) Soldier

(*c*) Queen termite
(*d*) King

Main features of termites are as follows:

- medium-sized, rather soft-bodied insects
- two pairs of very similar wings possessed by reproductives but shed after the mating flight
- antennae are moniliform, of moderate length
- mouth parts of generalised biting pattern
- cerci short, with many segments
- development hemimetabolous with only slight metamorphosis

Termites feed mostly on dead plant material and some are serious pests of timber buildings and other wooden structures in tropical countries. However, a few species attack growing plants (e.g. young cocoa). In other situations termites have a beneficial role in assisting with soil formation in the drier parts of the tropics. They carry vegetable material into the soil and break up the soil to permit aeration and penetration by water.

Answer the following questions:

1. Why termites are called white ants? ______________________
____
2. What changes occur in the life stage of a queen termite? ______________________________
3. How will you collect reproductives and when? ______________________________
4. Can termites be beneficial to man? Yes/No

   Give reasons for your answer. ______________________________
____
____
5. What is a termitarium? ______________________________
6. How termites can digest the cellulose from wood? ______________________________
____

ORDER MALLOPHAGA

Common name: biting lice, bird lice

All biting lice are ectoparasitic on birds for the most part with a few species infesting mammals such as sheep, cattle, dogs and horses. They do not attack pigs or man. The entire life cycle is spent on the host. Feeding is mainly on feathers, hair and other epidermal products but blood meals may sometimes be taken.

Examine a permanent slide of sheep body louse and observe the following features of Mallophaga:

- small, wingless insects (secondarily apterous)
- mouth parts of modified biting pattern

- antennae three-to-five jointed
- eyes reduced; no ocelli present
- thoracic spiracles ventrally placed
- tarsi with one or two joints
- abdominal cerci absent
- slight metamorphosis seen

Label the diagram in Fig. 7.6

Answer the following questions:

1. What is the impact of biting lice on poultry and cattle? ____________________
2. Why Mallophaga is not grouped under Apterygota? ____________________

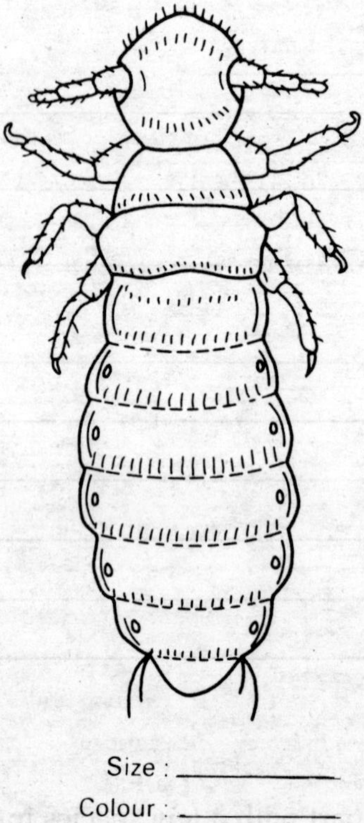

Size : ________

Colour : ________

Fig. 7.6:

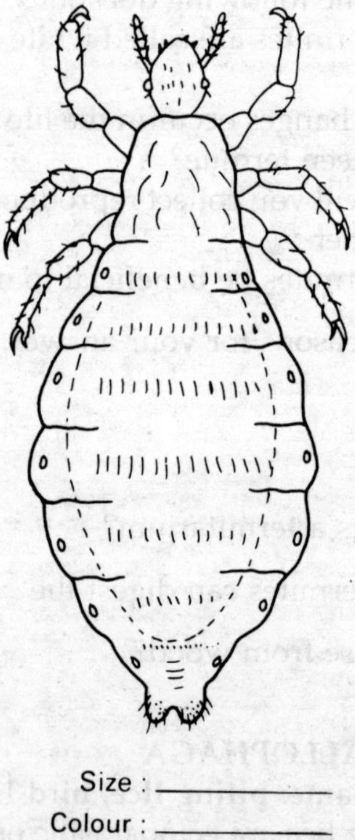

Size : ________

Colour : ________

Fig. 7.7:

ORDER **SIPHUNCULATA**

Common name : sucking lice

Examine a permanent preparation of sucking lice and study its characteristic features:

- small wingless insects
- body dorso-ventrally flattened

- eyes reduced or absent; ocelli absent
- antennae 3-5 jointed
- mouth parts modified for piercing and sucking, retractable.
- all three thoracic segments fused
- thoracic spiracles situated dorsally
- prominent legs with unjointed tarsi bearing large single claws
- cerci absent

Provide a heading to Fig. 7.7 and label the diagram.

The sucking lice are exclusively ectoparasites of mammals and unlike the biting lice, they are entirely blood sucking. When they attack domestic animals, there is loss of condition and poor growth, and fall in milk yield in dairy cattle. The human louse, *Pediculus humanus* attacks man and transmits typhus.

Answer the following questions:

1. Give three main differences between a biting and sucking lice. ______________________
2. Write the features which adapt this insect for the ectoparasitic habit. ______________________
3. Without wings, how this parasite is transferred around? ______________________
4. Give the economic importance of sucking lice to man. ______________________

ORDER HEMIPTERA

Common name: bugs—includes leafhoppers, cicadas, aphids, scale insects, mealy bugs, white flies

The Hemiptera are a very diverse group in appearance but are characterized by the invariable possession of piercing and sucking mouth parts of the same basic pattern. They are able to partake only of liquid food. Besides plant feeders some Hemiptera are predatory on other insects and others feed on blood of vertebrates.

The hemiptera are important plant pests and may damage plants in several ways:

1. Withdrawl of sap, weakening the plant
2. Injection of toxic saliva producing deformed or stunted plant parts
3. Growth of sooty mould on honeydew excreted by the insects
4. Transmission of plant viruses (applies to aphids and leafhoppers in particular)

The Hemiptera are divided into two distinct sub-orders:

(1) **Sub-order Heteroptera**: Forewings leathery at the base, membranous at apex. Folded flat over the body when at rest (Plant bugs).

(2) **Sub-order Homoptera:** Forewings of uniform texture throughout. Folded tent-like over the body when at rest (cicadas, leafhoppers, aphids, scale insects, mealybugs, white flies). Many species of Homoptera are wingless and degenerate.

SUB-ORDER : HETEROPTERA

Examine and label specimens (Figs. 7.8 & 7.9) of green vegetable bug and wheat bug and observe the following features which are found in both species:

- forewings of two textures (characteristic of all Heteroptera). Using a needle seeker, carefully lift one pair of wings to the side. Note that the basal part of the forewing is leathery in texture whereas the tip is membranous but that the hind wing is entirely membranous.
- body broadly flattened and shield like.
- mouth parts of typical piercing/sucking type.

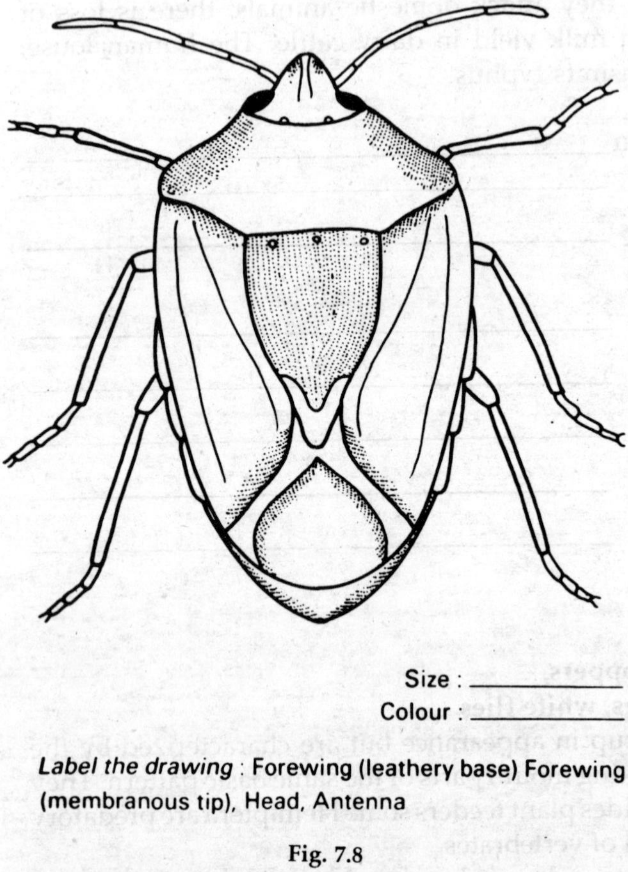

Size : ____________

Colour : ____________

Label the drawing : Forewing (leathery base) Forewing (membranous tip), Head, Antenna

Fig. 7.8

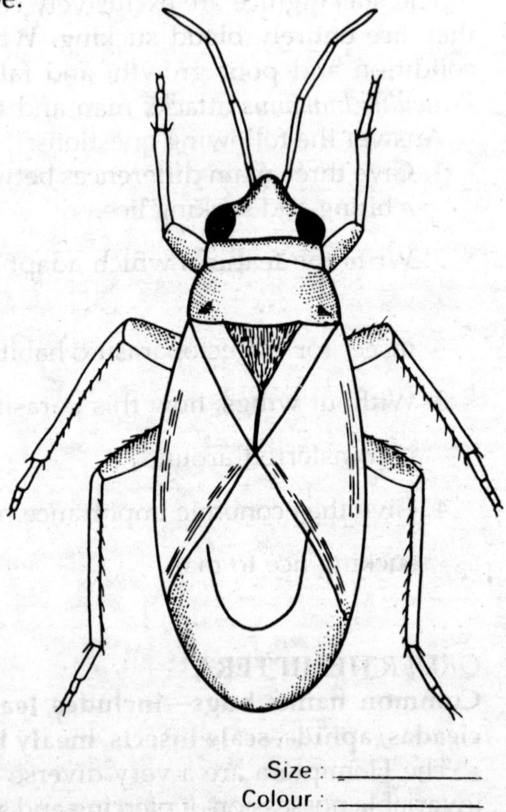

Size : ____________

Colour : ____________

Label the drawing : Forewing (leathery base), Forewing (membranous tip), Head, Antenna

Fig. 7.9:

SUB-ORDER : HOMOPTERA

Aphids

Aphids are an extremely important pest group, almost every kind of cultivated plant being subject to attack by one or more species. They are the main insect vectors of plant viruses in temperate climates.

Examine aphid specimens (live or preserved) and observe the following features:

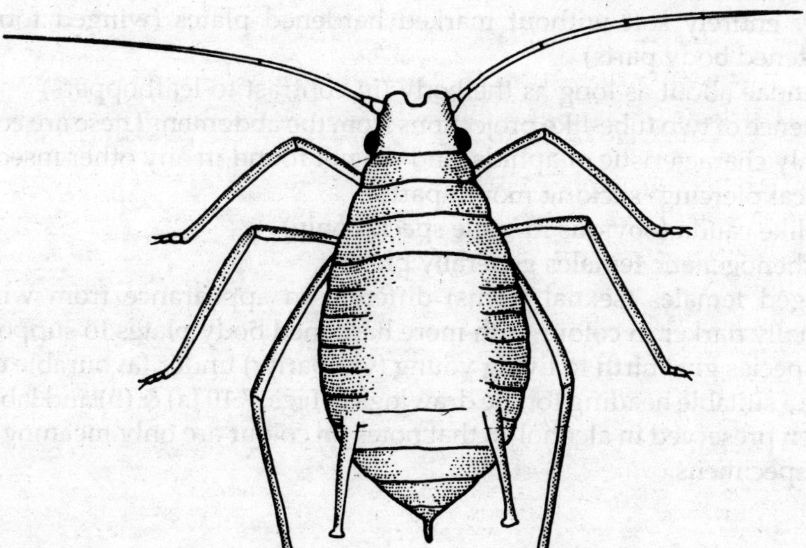

Size : ____________

Colour : ____________

Label the drawing : Antenna, Cornicles, Cauda

(a)

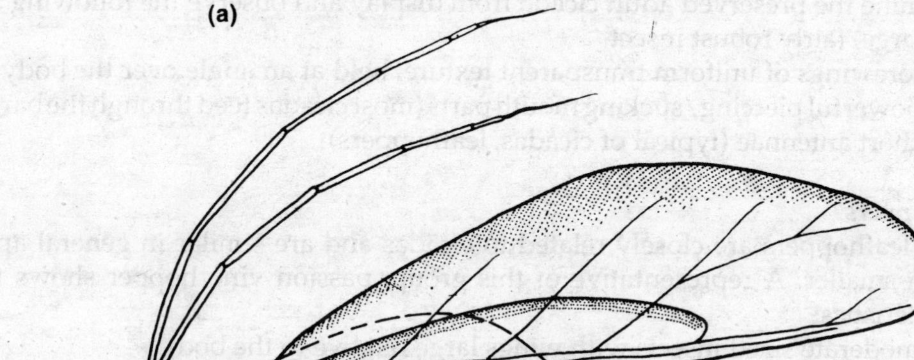

Size : ____________

Colour : ____________

Label the drawing : Antenna, Cornicles, Forewing, Hind wing

(b) **Fig. 7.10 (a & b)**

- body entirely soft without marked hardened plates (winged forms have more hardened body parts)
- antennae about as long as the body (in contrast to leafhoppers)
- presence of two tube-like projections from the abdomen. These are cornicles and are highly characteristic of aphids, and are not found in any other insects.
- typical piercing/sucking mouth parts
- tail-like cauda (obvious in some species only)
- parthenogenetic females generally present
- winged females (sexual forms) different in appearance from wingless females (usually darker in colour with more hardened body plates to support wings)
- all species give birth to living young (viviparity) under favourable conditions.

Provide a suitable heading for the drawings in Figs. 7.10 (a) & (b) and label. Aphids fade badly when preserved in alcohol so that notes on colour are only meaningful when made from live specimens.

Cicadas

The only significance of cicadas as pests is from the egg incisions made by the female in small branches of various trees and shrubs. These can result in weakening and breakage. The nymphs are subterranean and feed on the roots of various plants. They take several years to develop.

Examine the preserved adult cicada from display and observe the following features:

- large, fairly robust insect
- forewings of uniform transparent texture, held at an angle over the body
- powerful piercing/sucking mouth parts (most cicadas feed through the bark of trees)
- short antennae (typical of cicadas, leafhoppers)

Leafhoppers

The leafhoppers are closely related to cicadas and are similar in general appearance though smaller. A representative of this group, passion vine hopper shows following characteristics:

- moderate sized insects with wings large relative to the body
- forewings mottled in appearance (not in all leafhoppers)
- typical piercing/sucking mouth parts
- short antennae

Some species of leafhoppers are very important pests directly (e.g. *Pyrilla perpusilla* on rice) or act as virus vectors. Most species hop vigorously when disturbed but also have the ability to fly. Provide a heading for Fig. 7.11 (a) and label them. Study Fig. 7.11 (b) for a typical nymph of a leafhopper.

Scale insects

Scale insects are so named because most stages are apparently inert objects with a waxy scale-like covering firmly attached to plant surfaces. The insect itself is a featureless sac without legs, antennae or other obvious features. The outer covering is a waxy scale secreted by the insect and apparently separate from it or as part of the body.

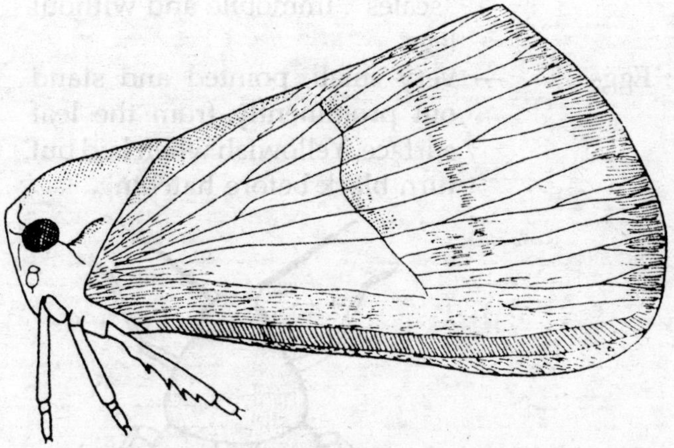

Size : ____________________

Colour : ____________________

Label the drawing : Antenna, Forewing

(a)

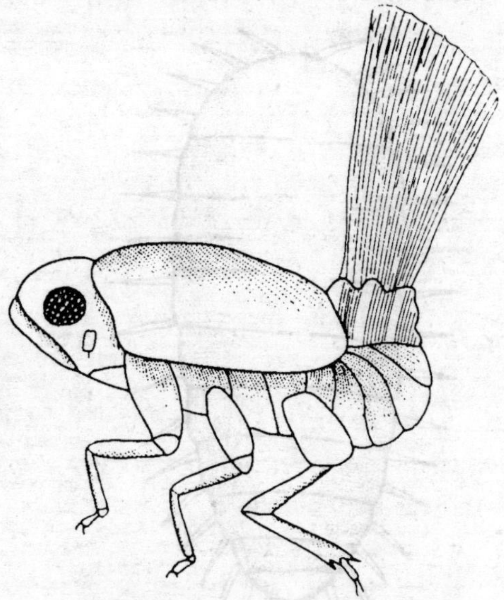

Size : ____________________

Colour : ____________________

Label the drawing : Tuft of waxy tail filaments

(b)

Fig. 7.11 (a), (b)

Piercing/sucking mouth parts are present in the form of very long threadlike stylets permanently embedded in the plant tissue. The outer sheath of the proboscis is absent. Separate male and female scales occur in some species; others are parthenogenetic with females only. Adult males, when they occur, are tiny winged fully formed insects. The first stage nymphs of scale insects possess legs and can move actively. They are referred to as crawlers and are the means of dispersal.

Example is lac insect, *Laccifer lacca* which has proved beneficial to man (otherwise scale insects are generally injurious to plants).

Mealybugs

Mealybugs are closely related to the scale insects and show following features:

- insects soft-bodied and covered with flaky white wax (though no scale formed)
- three pairs of legs present but small and concealed on the underside
- outgrowths of waxy filaments around the margins of the body and at the tail end giving the appearance of many legs.

Provide a heading to Fig. 7.12.

White flies

White flies show different stages (Fig. 7.13 a and b):

Adults	—	small, chalky, white moth-like insects, can fly.
Nymphs	—	flattened, semi-transparent "scales". Immobile and without legs.
Eggs	—	very small, pointed and stand out prominently from the leaf surface. Yellowish when laid but turn black before hatching.

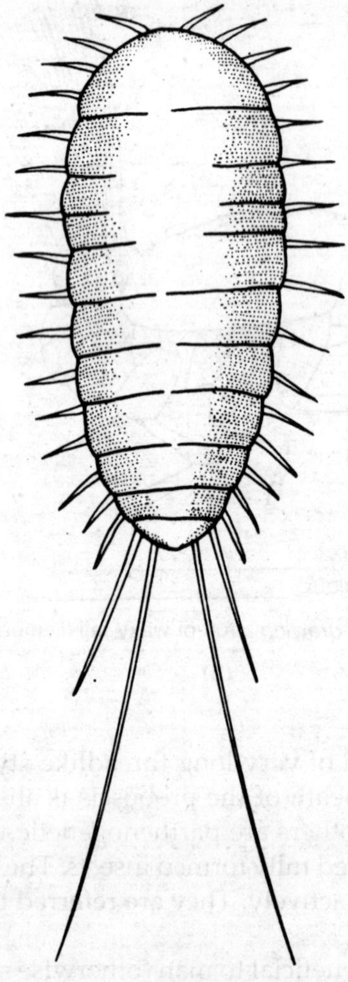

Size : ____________

Colour : ____________

Label the drawing : Waxy filaments

Fig. 7.12

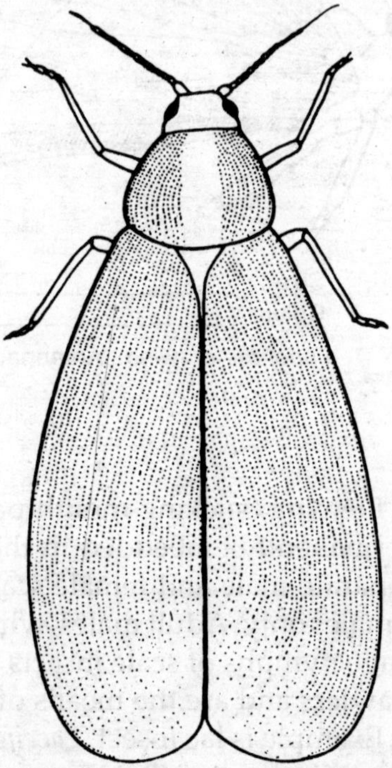

Size : ____________

Colour : ____________

Label the drawing : Forewings

Fig. 7.13 (a)

Label Fig. 7.13 (a) and make a drawing of the scale stage [Fig. 7.13 (b)].

Answer the following questions:

1. Does the provided specimen belong to the Hemiptera? Yes/No ____________

   Give reasons for your answer. ____________

2. To which sub-order of Hemiptera does specimen A belong? ____________

3. How would you distinguish an aphid from other small soft-bodied insect? ______________________
4. List three types of plant injury that may be caused by aphids. ______________________
5. Why are low numbers of aphids significant in some crops? ______________________
6. What is honeydew? ______________________
7. Name two types of plant injury caused by wheat bug. ______________________
8. How do scale insects disperse? ______________________

ORDER **THYSANOPTERA**

Common name: thrips

The following features are characteristic of all Thysanoptera:

- two pairs of strap-like wings fringed with long hairs present in adults.
- cone-shaped piercing mouth parts, tucked well under the head
- antennae with 6-10 joints
- tarsi with only 1 or 2 joints and terminating in a swollen vesicle instead of claws
- nymphs of similar general appearance to adults but without wings
- droplets of honeydew can sometimes be seen at the tip of the abdomen of live thrips (especially nymphs)

Feeding in thrips is by piercing of superficial plant cells and sucking out of the contents. This often results in a silvery appearance on leaves of damaged plants. Distortion of attacked plant parts and reddening of leaves are other common symptoms. Sometimes specks of dark brown varnish-like excreta on underside of leaves are also seen.

Provide a heading for Fig. 7.14 and indicate the approximate size and colouration.

ORDER **NEUROPTERA**

Common name: lacewings

Lacewings are important in that they are all predatory on other insects, particularly aphids. They are holometabolous. Examine specimens of a larva and an adult and observe the following features:

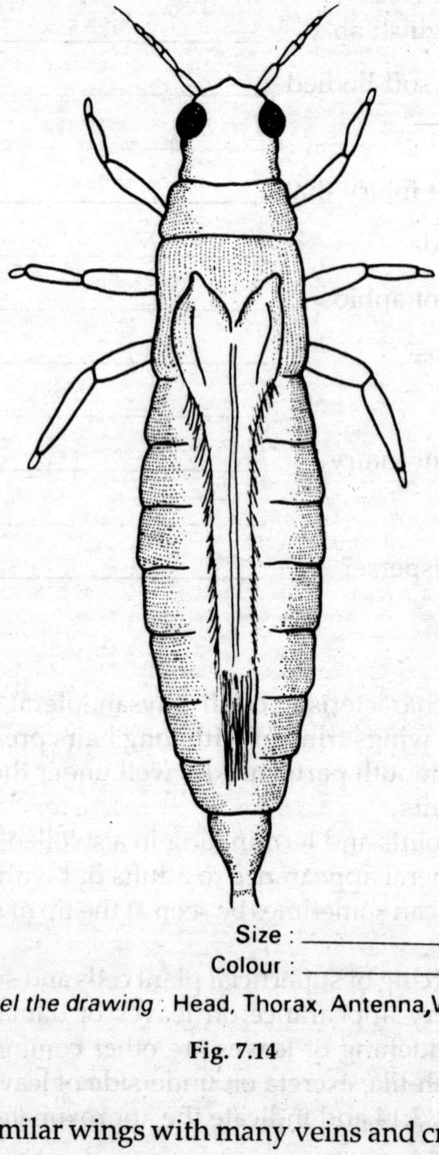

Size : ___________

Colour : ___________

Label the drawing : Head, Thorax, Antenna, Wings

Fig. 7.14

Adult

- two pairs of large similar wings with many veins and cross-veins forming a net-like pattern
- antennae long and many jointed

Larva

- body tapered towards both ends
- antennae much shorter than those of the adult
- very prominent pointed mandibles so that they can pierce their prey followed by the body contents being sucked out.

Provide suitable headings for the Fig. 7.15 a and b.

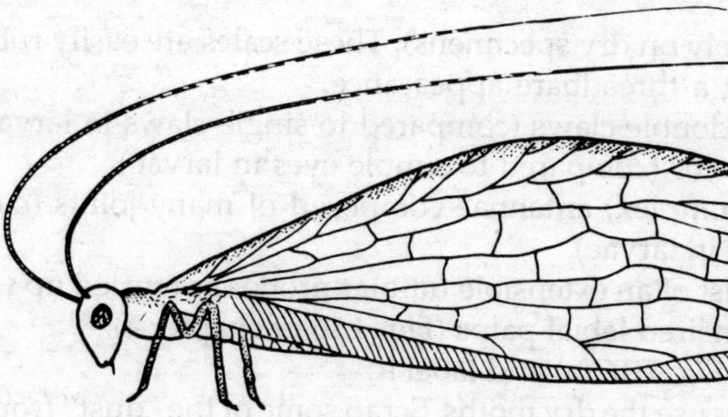

Size : ____________
Colour : ____________

Label the drawing : Wings (two pairs), Antenna

(a)

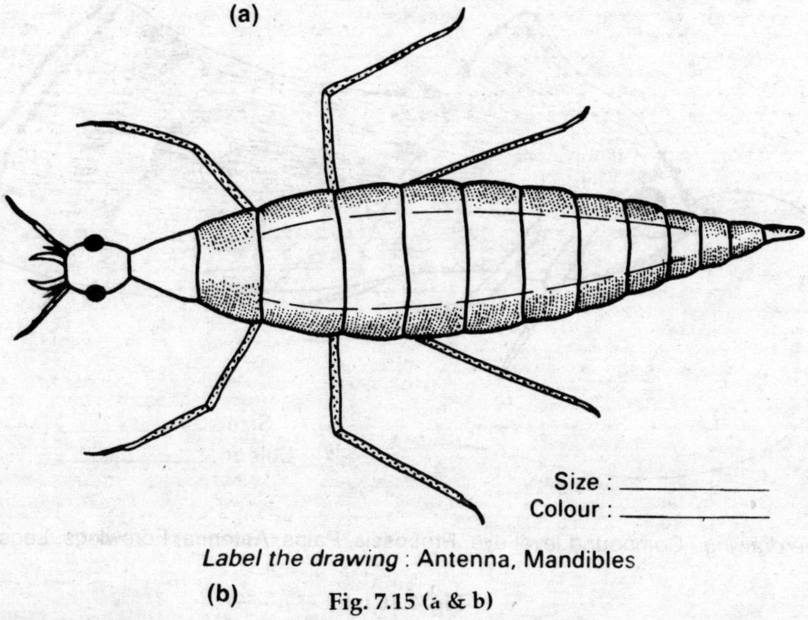

Size : ____________
Colour : ____________

Label the drawing : Antenna, Mandibles

(b)

Fig. 7.15 (a & b)

ORDER **LEPIDOPTERA**

Common name: Butterflies and moths

This is a large and important order of insects containing many pest species. Damage is caused by the larvae (caterpillars) which possess strong biting mouth parts. The adults are liquid feeders (on nectar primarily) and are normally incapable of injuring plants. The adults of some species (e.g. porina moth) do not feed and have reduced mouth parts.

We shall study the adult, pupa and larva of a noctuid moth (a common family of Lepidoptera that includes armyworms and cutworms).

Adult

Observe the following features which are characteristic of all adult butterflies and moths (with a few minor exceptions):

- body covered in a dense "fur" of hairs
- colour and pattern of the wings is provided by innumerable minute scales. (To be

examined separately on dry specimens). These scales are easily rubbed off in older specimens, giving a threadbare appearance.

- legs terminate in double claws (compared to single claws in larvae)
- large compound eyes (compared to simple eyes in larvae)
- long (and often complex) antennae composed of many joints (compared to very reduced antennae in larvae)
- mouth parts consist of an extensible tubular proboscis (curled up when at rest) and a pair of densely haired labial palps (Fig. 4.16 of chapter 4).

Provide a heading to Fig. 7.16 (a) and label it.

To examine *wing scales* use the dry moths. Scrap some of the "dust" from the wings onto a microscope slide. Do not wet it nor apply a coverslip. Examine under the compound microscope.

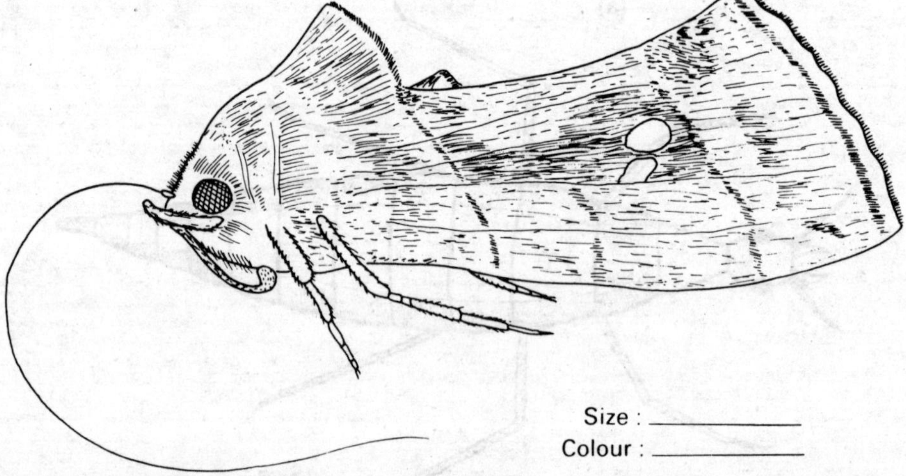

Size : ____________

Colour : ____________

Label the drawing : Compound level eye, Proboscis, Palps, Antenna, Forewings, Legs (3 pairs)

Fig. 7.16 (a)

Larva

A Lepidoptera larva has already been examined in chapter 4 as a second example of the main external features of insect anatomy (Fig. 4.26 a). Mouth parts of a caterpillar (biting pattern) have similarly been examined (Fig. 4.10 of chapter 4).

However, larvae of Lepidoptera may easily be confused with larvae of sawflies (order Hymenoptera) which in most species are very caterpillar-like and feed in a similar manner. Points of distinction between lepidopterous and sawfly larvae are:

TABLE 7.2: Distinction between lepidopterous and sawfly larva

Feature	*Larvae of Lepidoptera*	*Larvae of Sawflies (Hymenoptera*)*	*Specimen 1 Order:*	*Specimen 2 Order:*
false legs	5 pairs or less; none on second abdominal segment	6-10 pairs; always one pair on second abdominal segment		
ocelli	several on each side of head	one only on each side of head		
crotchets	crotchets present as patterns of spines at the extremity of each false leg	crotchets absent from false legs		

*Larvae of other Hymenoptera (ants, bees, wasps) are legless and quite different from those of sawflies.

Complete the table for specimens 1 and 2.

Many lepidopterous larvae spin silken threads which they use to construct cocoons for pupation (e.g. in silk moth, *Bombyx mori* these silken threads "excreted" by larvae to form the puparium are commercially exploited to produce the silk). The silk is produced by large internal glands which open through a hollow spine (spinneret) on the labium of the mouthparts.

Provide a heading to Fig. 7.16 (b) of a caterpillar and label it.

Pupa

The pupae of butterflies and moths have all the appendages (wings, legs, antennae) smoothly enclosed though their position can easily be seen by external markings on the pupa. The pupae of moths are often enclosed in a silken cocoon. Provide a heading to Fig. 7.16 (c) and label it.

Sometimes small moths may be confused with caddis flies (order Trichoptera) which are small night flying insects that are moth-like in general appearance. Caddis flies are attracted to light, as are many moths, and are very common near flowing water (the larvae are aquatic). The main points of distinction are :

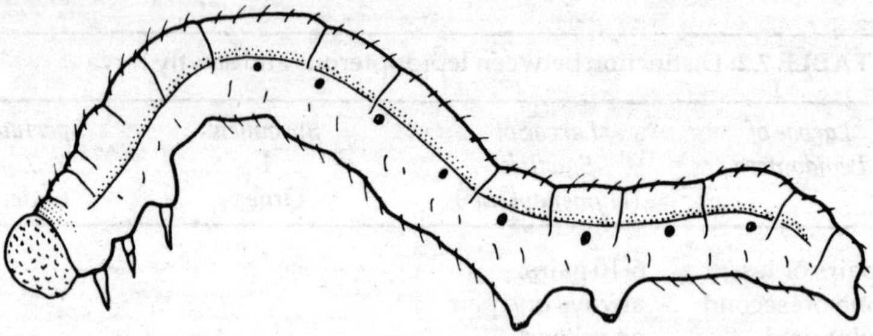

Size : ____________

Colour : ____________

Label the drawing : Head, Jointed legs, false legs, Spiracles

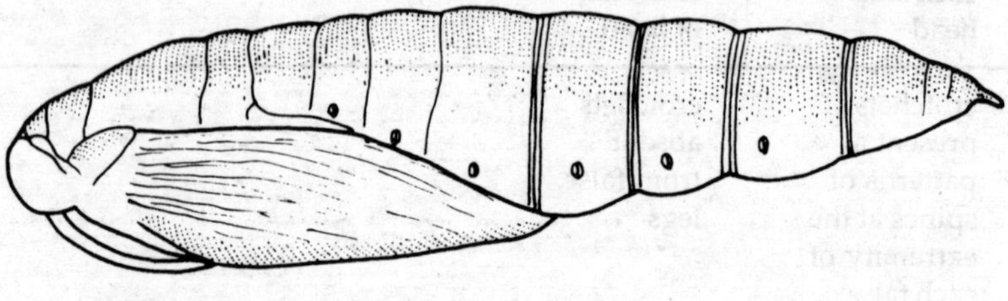

Size : ____________

Colour : ____________

Label the drawing : Head, Jointed legs, False legs, Spiracles

Fig. 7.16 (b and c)

TABLE 7.3. Distinction between a moth and caddisfly

Feature	*moth (Order Lepidoptera)*	*caddisfly (Order Trichoptera)*	*Specimen 3 Order:*	*Specimen 4 Order:*
wings	wings covered with scales (flattened, plate-like)	wings clothed with hairs (tapering, pointed)		
mouth parts	coiled tubular mouth parts with prominent palps (mouth parts not developed in some species)	mouth parts never developed though palps may be prominent		

Complete the table for specimens 3 and 4.

Answer the following questions:

1. What is specimen A ? ____________

   Give reasons for your answer. ____________

2. Describe in general terms the nature of plant injury caused by larvae of Lepidoptera. ____
3. Do adult butterflies or moths injure plants? ____
4. Name two lepidopterous pests that commonly attack brassicas. ____
5. Where would you expect to find eggs of white butterfly? ____
6. Give common names of the following 2 moth pests of stored grain:
   Sitotroga cerealella ____________________
   Plodia interpunctella ____________________

ORDER **COLEOPTERA**

Common name: beetles, weevils

This is the largest order of insects with about a quarter of a million species. It contains many important pests of crops, stored products and timber. All types of plant feeding behaviour may be found; root feeding, open leaf feeding, concealed feeding in leaves and shoots, and wood boring. Some groups of beetles are predatory on other insects, and the adults and larvae of Ladybird beetles in particular are important predators of aphids, and some other pests.

Two examples of Coleoptera will be exmained : grass grub beetle and grain weevil. The weevils are an important sub-group (family Curculionidae) within the beetles and there are many species of economic importance. In all cases the larvae of weevils are legless whereas most other beetle larvae possess well developed legs.

***Costelytra zealandica* (grass grub beetle)**

Adult: Examine the grass grub beetle and observe the following features:

- heavily armoured exoskeleton over most of the body. However, areas beneath the wing cases are soft.
- forewings in the form of hard horny cases (**elytra**) which enclose and protect the hind wings and the dorsal surface of the abdomen.
- lift the elytra and note the membranous hind wings beneath (in a folded condition). Beetles can fly, though are not strong fliers.
- mouth parts of a biting pattern but the mandibles are difficult to see clearly.

Provide a heading to Fig. 7.17 (a) and label the various parts.

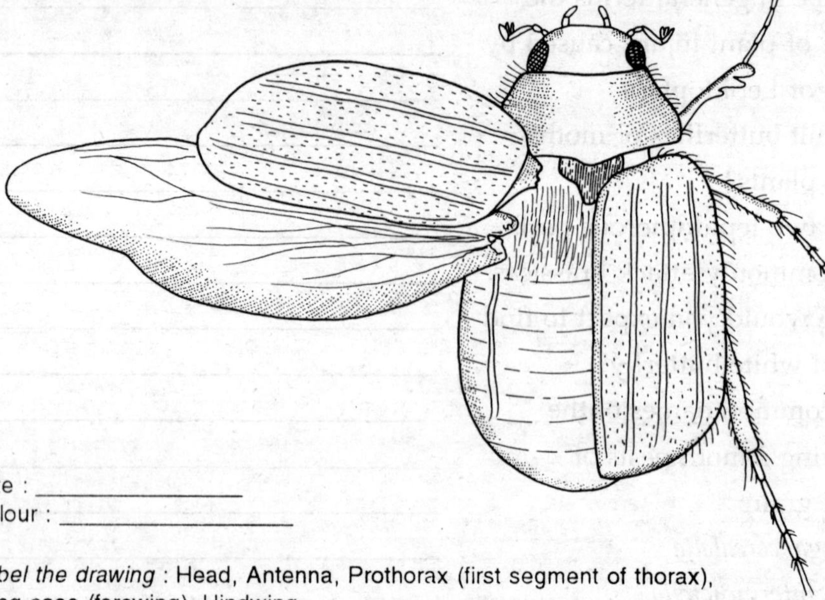

Size : ____________

Colour : ____________

Label the drawing : Head, Antenna, Prothorax (first segment of thorax), Wing case (forewing), Hindwing

Fig. 7.17 (a)

Larva: Observe the following points in a grass grub larva :

- generally soft-bodied except for hard head
- three pairs of jointed legs arising from the thorax terminate in single claws

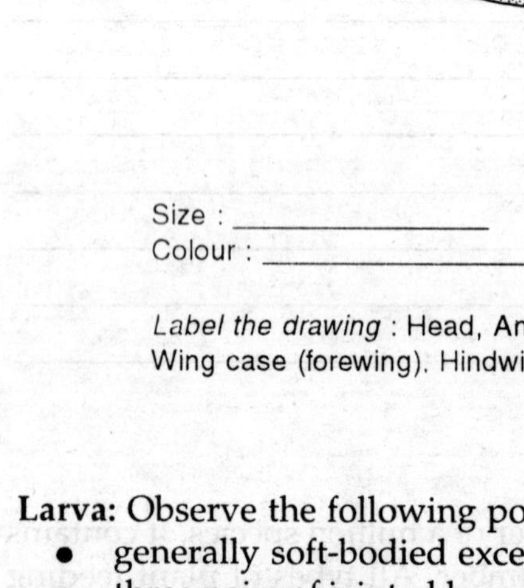

Size : ____________

Colour : ____________

Label the drawing : Head, Antenna, Mandible, Legs, Thorax

(b)

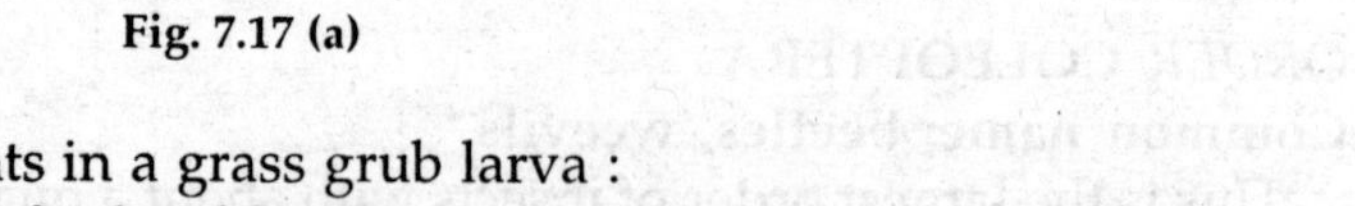

Size : ____________

Colour : ____________

(c)

Fig. 7.17 (b & c)

- antennae fairly short (but more prominent than in a lepidopterous larva)
- no compound eyes and no ocelli (not required in a subterranean insect)
- mouth parts of biting/chewing pattern but with pointed mandibles (black towards the tip)
- a horse-shoe shaped ring of dark spines at the hind end near the anus, which distinguishes the larva of *Costelytra* from other scarabs.

Provide a heading to Fig. 7.17 (b) and label it.

Pupa: The pupae of beetles (unlike those of butterflies and moths) have all the appendages (wings, legs, etc.) separately encased, i.e., exarate type (Fig. 7.17 c). The pupa usually occurs within a crudely formed cell, often in the soil.

***Sitophilus granarius* (Grain weevil)**

Adult: Examine the adult *Sitophilus* collected from infested grains:

- lower front portion of the head drawn out to form a snout or rostrum (typical of all weevils but length varies according to species)
- mouth parts (of biting pattern) located at the tip of the rostrum, but details are difficult to see as the mouthparts tend to be enclosed.
- antennae 5-6 jointed, the last segment being large and club-shaped. The basal joint fits into a special groove along the side of the rostrum.
- this weevil can fly but some weevils are flightless because of jointed elytra.

Provide a heading to Fig. 7.18 and label different parts.

Larva: Examine the grain weevil larva and observe the following features:

- generally soft-bodied except for the head. Head is dark coloured and tends to be partially retracted into the thorax
- no compound eyes and no ocelli
- antennae very reduced
- prominent biting mandibles which are black towards the tip. The rest of the mouth parts are reduced with only very small palps present.

The larvae of weevils are very similar to those of many Hymenoptera (ants, bees and wasps) in appearance. Make a drawing of *Sitophilus* larva and label head, mandibles in it. Indicate size and colour.

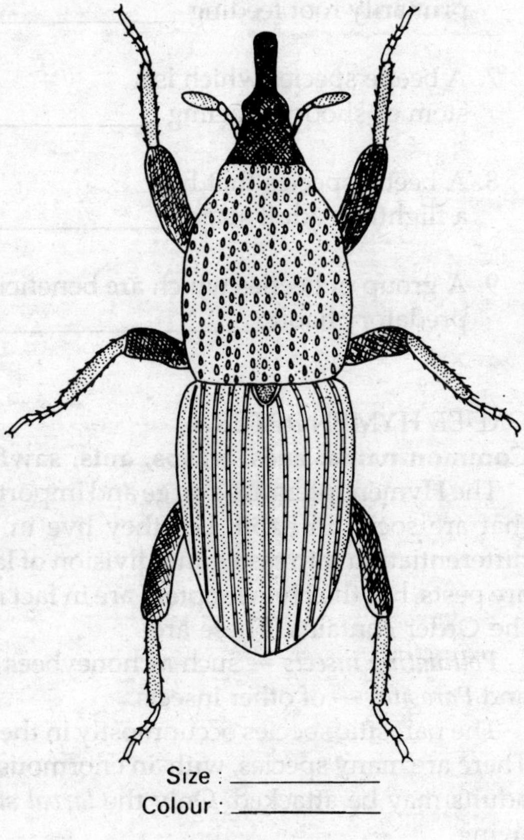

Size : ________

Colour : ________

Fig. 7.18

Answer the following questions:

1. Does specimen A belong in the Coleoptera? Give reasons for your answer. ____________ Yes/No
2. Is specimen B the larva of beetle ? ____________ Yes/No
   Give reasons for your answer. ____________

   ____

Give one example of each of the following:

3. A beetle species which is a pest primarily as a larva ____________
4. A beetle species which is a pest primarily as an adult ____________
5. A beetle species which is a pest both as a larva and as an adult ____________
6. A beetle species which is primarily root feeding ____________
7. A beetle species which is a stem or shoot tunneling ____________
8. A beetle species that has a flightless adult stage ____________
9. A group of beetles which are beneficial because of their predatory habit ____________

ORDER HYMENOPTERA

Common name: bees, wasps, ants, sawflies

The Hymenoptera are a large and important group of insects. They contain many species that are social in habit, i.e., they live in complex highly organized communities with differentiation into castes and division of labour. Some species of ants, wasps and sawflies are pests, but the Hymenoptera are in fact more important for the beneficial species which the Order contains. These are:

Pollinating insects — such as honeybees and other bees and *Parasites* — of other insects.

The parasitic species occur mostly in the subroders Ichneumonoidea and Chalcidoidea. There are many species, with an enormous range of size. All stages of insects from eggs to adults may be attacked. Only the *larval stages are parasitic,* the adults always being free living.

The main features of adult and larval stages of Hymenoptera are best seen in a larger species such as the German wasp. The adult of a parasitic wasp (Ichneumon wasp) will also be examined.

German Wasp

Adult: Examine the adult German wasp and observe the following features which are typical of adult Hymenoptera:

- two pairs of membranous wings, the hind wings being considerably smaller. The fairly simple venation with a dark blotch (stigma) on the front margin of the forewing is characteristic of Hymenoptera. In small parasitic species the wing veins are greatly reduced. The fore and hind-wings are normally hooked together and operate as one.
- a pronounced constriction (*"waist"*) apparently between the thorax and abdomen (not present in sawflies). This narrow waist is in fact formed from the second segment of the abdomen. The first segment is incorporated into the thorax and is referred to as the *propodeum*.
- determine whether your specimen is:
  - *(i)* a *worker* (sterile female), abdomen rather globular and coming to a point where the sting is located, or
  - *(ii)* a *drone* (male), abdomen rather elongate and more blunt ended, or
  - *(iii)* a *queen* (fertile female), body form as for a worker but considerably larger.

If your specimen is a worker, examine the sting at the extreme tip of the abdomen. It tends to be retracted inside but can usually be partially extended by gently squeezing the abdomen. The sting is formed from the ovipositor (refer to chapter 4, Fig. 4.24 c).

- mouthparts are of biting/chewing type. Identify the mandibles and palps (two pairs). In bees, however, the mouth parts are modified for lapping rather than for biting (discussed earlier, chapter 4, Fig. 4.15).

Provide a heading to the Fig. 7.19 (a) and label it.

Larva: Examine the larva of the German wasp and observe the following features typical of a larval Hymenoptera:

- soft bodied but with a hardened head which may be somewhat retracted into the thorax (head prominent in sawfly larvae)
- legless (larvae of sawflies possess legs)
- no compound eyes or ocelli present (ocelli present in sawfly larvae)
- antennae absent (very reduced in sawfly larvae)
- mouth parts of biting type. Mandibles are clearly visible (black towards the tips) but palps reduced to tiny papillae.

A larva of a sawfly, which is caterpillar-like, has already been examined with the lepidopteron larva.

Provide a heading to Fig. 7.19 (b) and indicate its size.

Pupa: The pupae of Hymenoptera, like those of beetles, are of exarate type and have all the appendages separately encased.

Label the drawing of a wasp pupa in Fig. 7.19 (c).

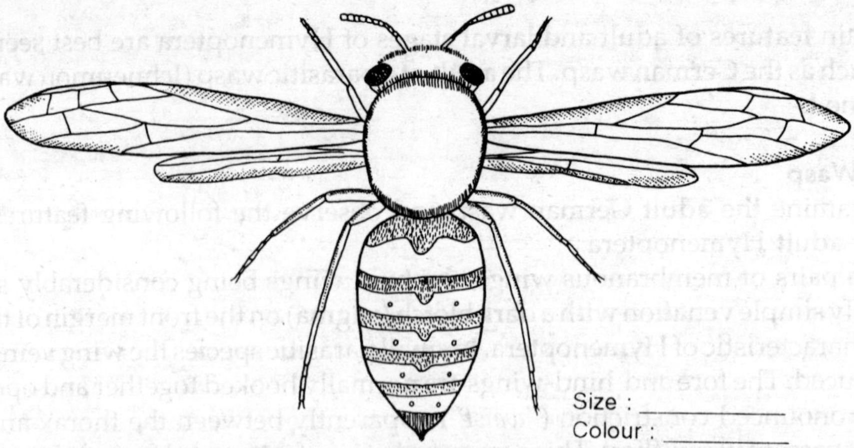

Size : ____________

Colour : ____________

Label the drawing : Head, Antenna, Abdomen, "Waist" Forewing, Hindwing, Stigma (on forewing), Tip of sting (ovipositor)

(a)

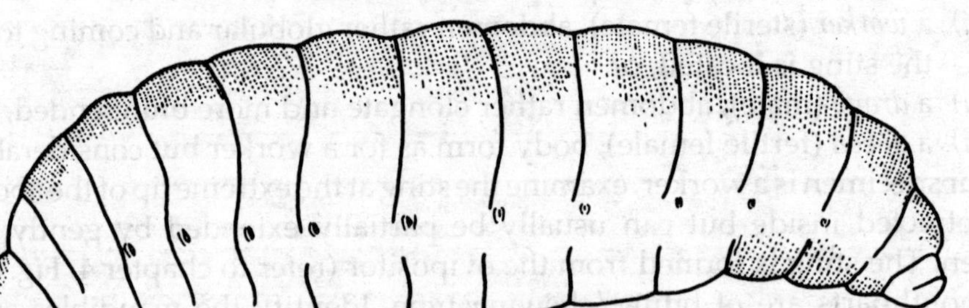

Size : ____________

Colour : ____________

Label the drawing **: Head, Mouth parts, Spiracles**

(b)

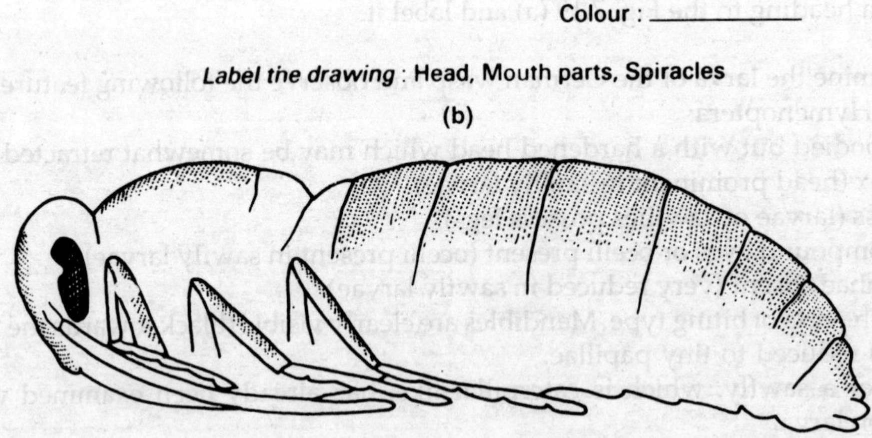

Size : ____________

Colour : ____________

Label the drawing : Head, Thorax, Abdomen, Antenna, Legs

(c)

Fig. 7.19 (a, b, c)

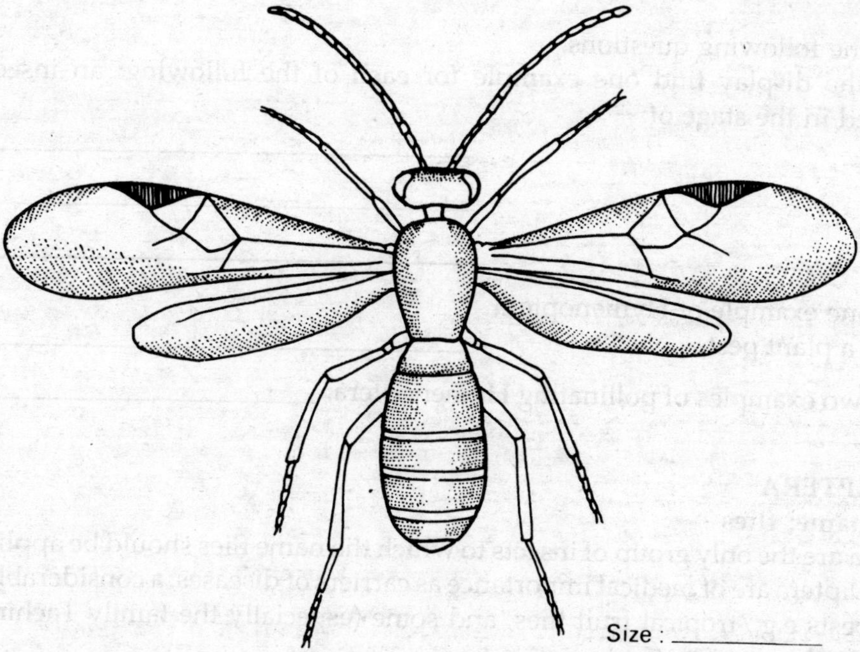

Size : ___________

Colour : ___________

Label the drawing : Head, Antenna, "Waist", Forewing, Stigma, Hindwing

Fig. 7.20

Ichneumon wasp (*Apanteles glomeratus)*

Adult: Observe the following features and note the general similarity to the German wasp:

- two pairs of membranous wings, with a dark stigma on the front margin of the forewing
- a distinct waist between thorax and abdomen
- biting mouth parts with prominent palps
- no sting but an ovipositor present (eggs are laid into insects which are attacked)

Provide a heading to Fig. 7.20 of an hymenopterous parasite adult and label it.

Larva: The larva of *Apanteles* (parasitic wasp) is a parasite of white butterfly larva. It has no legs, antennae or eyes just like the German wasp larva. This degenerate nature of most larval Hymenoptera (except for sawflies) is due to the fact they are either fed and tended within a nest (ants and many wasps) or are surrounded by a rich food medium in the form of host insect body tissue in the parasitic forms. Elaborate sense-organs and legs are therefore not required.

Note that not all parasites are in the order Hymenoptera. Adult insects are rarely parasitized and that parasites may themselves be attacked by other parasites (secondary or hyperparasites).

Answer the following questions:

1. From the display find one example for each of the following: an insect that is parasitised in the stage of:

   egg ____________________

   larva ____________________

   pupa ____________________

2. Give one example of Hymenoptera that is a plant pest. ____________________

3. Give two examples of pollinating Hymenoptera. ____________________

ORDER DIPTERA

Common name: flies

The Diptera are the only group of insects to which the name flies should be applied. Many species of Diptera are of medical importance as carriers of diseases, a considerable number are plant pests e.g., tropical fruit flies, and some (especially the family Tachinidae) are parasites of other insects.

We will examine the life stages of the housefly, *Musca domestica* to illustrate the main characteristics of the order:

Adult of housefly

- one pair of wings (forewings)
- hidwings converted to specialised balancing organs called *halteres*. These are like tiny whitish drumsticks, and are best seen in view from the side or beneath. They are often obscured by the membranous flaps at the base of the forewings which act as protective covers (Fig. 7.21 (b)
- very large eyes occupying a high percentage of the surface of the head.
- antennae short with a large fleshy lobe from which arises a feathery bristle, *arista*.
- soft, extensible mouth parts of the sponging type which can only deal with liquid or semi-liquid food. This may be assisted by partial external digestion (Fig. 4.17, chapter 4).

Provide a heading to Fig. 7.21 (a) and indicate its size and colour.

Larva: Examine the housefly larva and observe the following features:

- legless, soft-bodied maggot, white in colour
- tapers towards the head end but without any distinct head
- slightly raised ridges at each body segment provide some grip for locomotion
- two pairs of spiracles only; the rear pair are obvious as two dark spots on the extreme hind end; the anterior pair are on the second segment back from the head.
- mouth parts in the form of a rasping hook (already discussed in an earlier chapter).

Make a drawing of the housefly larva in your notebook and label the following : head end, hind end, ridges at margins of each segment, posterior spiracles, anterior spiracles, mouth hook.

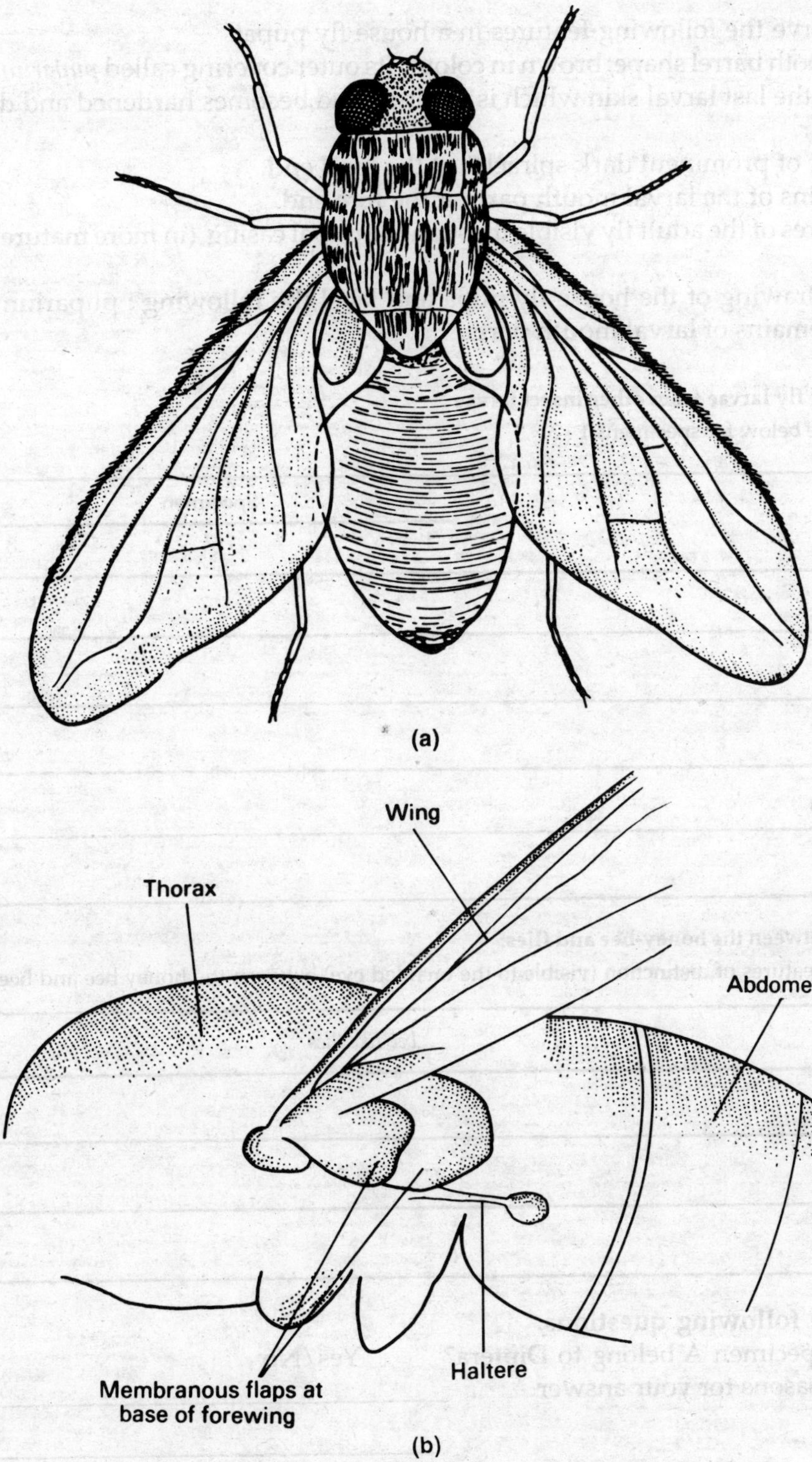
(a)
Wing
Thorax
Abdomen
Membranous flaps at base of forewing
Haltere
(b)

Fig. 7.21 (a & b)

Pupa: Observe the following features in a house fly pupa:

- a smooth barrel shape; brown in colour. Its outer covering called *puparium* is formed from the last larval skin which is retained and becomes hardened and darkened in colour.
- a pair of prominent dark spiracles at the hind end.
- remains of the larval mouth parts at the fore end.
- features of the adult fly visible through the pupal casing (in more mature specimens only)

Make a drawing of the house fly pupa and label the following : puparium, posterior spiracles, remains of larval mouth parts.

Distinction of fly larvae from other insect larvae

Fill in the table below for specimens 1 - 3.

Feature	Specimen		
Yes/No	1	2	3
distinct head			
true legs			
false legs			
mouth hooks			
fly larva			

Distinction between the honey-bee and flies:

Record three features of distinction (visible to the unaided eye) between the honey bee and bee-like flies.

honey bee	*bee-like flies*
1.	
2.	
3.	

Answer the following questions:

1. Does specimen A belong to Diptera? Yes/No
   Give reasons for your answer. ______________________
   ____

2. What is characteristic of the family Tachinidae of Diptera ? ______

3. Give two examples of fruit flies which are serious pests of fruits. ______

4. Besides house flies, which other genera act as carriers of diseases in man? ______

ORDER SIPHONAPTERA

Common name: fleas

Take a permanent slide of rat flea, *Xenopsylla cheopis* and observe under a compound microscope for the following features:

- small, wingless, hard bodied insects
- body laterally compressed
- adults ectoparasitic, larvae free living
- compound eyes absent, but one pair of ocelli usually present.
- antennae short, thickened and three-jointed
- mouth parts modified for piercing and sucking
- legs adapted for jumping, with enlarged coxae
- all three thoracic segments free
- no abdominal cerci and ovipositor
- a number of large spines present on the head and thorax known as combs or ctenidia
- development holometabolous with legless maggot-like larvae

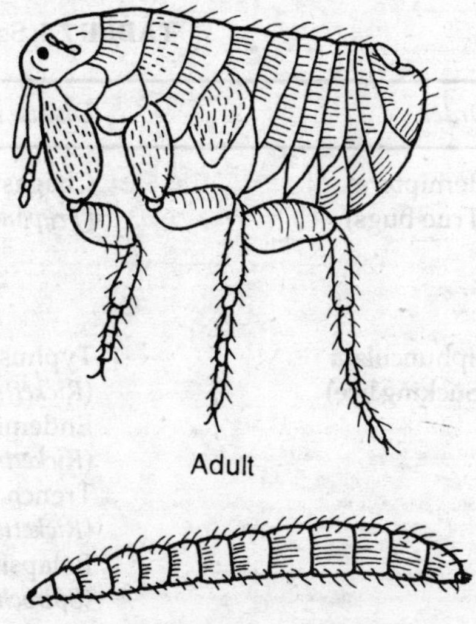

Fig. 7.22

Label the two life stages of a flea in Fig. 7.22.

The adults of all species of fleas are ectoparasitic and blood sucking on higher animals whereas the larvae develop free from the host and feed on organic debris. Fleas are less host specific than lice so that fleas from rats for instance, can transfer readily to humans, an important feature in relation to disease transmission. Adult fleas are known to transmit bubonic plague, murine typhus and certain salmonellas to man.

Answer the following questions:

1. What are the differences in the morphology and life cycle between a sucking lice and a flea? ______

2. Which organism is responsible for causing bubonic plague in man? ________________

3. Are the larvae of fleas parasitic? Yes/No
Give reasons for your answer. ________________

4. What are ctenidia and where they are located in a flea? ________________

5. Fleas are "facultative" ectoparasites. Explain. ________________

TABLE 7.4: Some Human Diseases Transmitted by Insects

Order	*Disease and Parasite*	*Transmitter Insect*
Hemiptera (True bugs)	Chagas' Diesease (*Trypanosoma cruzi*)	*Triatoma infestans, T. megista, T. chagasi, Rhodnius prolixus,* and others
Siphunculata (Sucking lice)	Typhus fever (*Rickettsia prowazeki*)	*Pediculus humanus capititis & P. h. corporis*
	Endemic typhus (*Rickettsia moseri*)	*Pediculus humanus*
	Trench fever (*Rickettsia quintana*)	*Pediculus humanus*
	Relapsing fever (*Spirochaeta recurrentis*)	*Pediculus humanus*
Diptera (The flies)	Kala Azar (*Leishmania donovani*)	*Phlebotomus argentipes, P. major, P. perniciosus*
	Carrion's disease or Oroya fever (*Bartonella bacilliformis*)	*Phlebotomus* spp.
	Oriental sore (*Leishmania tropica*)	*Phlebotomus papatasii*
	Espundia (*Leishmania brasiliensis*)	*Phlebotomus intermedius*
Diptera	Malaria (*Plasmodium vivax, P. malariae, P. ovale, P. falciparum*)	*Anopheles* spp. (more than 32 species)
	Yellow fever (Virus)	*Aedes aegypti* (and 16 other sp.)
	Dengue fever (?)	*Ae. aegypti, Ae. albopictus, Armigeres obturbans*

Order	*Disease and Parasite*	*Transmitter Insect*
	Filariasis (*Wuchereria bancrofti*)	*Aedes* spp. (8 species) *Anopheles* spp. (15 species) *Culex* spp. (3 species) *Mansonia* spp. (2 species)
	Filariasis (*Filaria malayi*)	*Mansonia* spp. (5 species) *Anopheles* spp. (2 species) *Culex* spp. (2 species)
	Myiasis (*Dermatobia hominis*)	*Culex* spp. *Psorophora* spp. *Goeldia* spp.
	Encephalitis	*Culex tarsalis, C. pipiens*
	Onchocerciasis (*Onchocerca volvulus*)	*Simulium damnosum, S. metallicum, S. callidum*
	Filariasis (*Acanthocheilonema perstans)*	*Culicoides austeni, C. grahami*
	Anthrax (Malignant pustule) (*Bacillus anthracis*)	*Tabanus* spp.
	Trypanosomiasis (Various trypanosomes)	*Tabanus* spp.
	Sleeping sickness (*Trypanosoma gambiense*)	*Glossina palpalis, G. morsitans*
	Sleeping sickness (*Trypanosoma rhodesiense*)	*Glossina morsitans, G. brevipalpis*
	Typhoid, dysentery, Cholera, anthrax, conjunctivitis, etc.	*Musca domestica*
	Myiasis	Larvae of *Sarcophaga, Calliphora, Phormia, Musca, Fannia, Hypoderma, Gasterophilus* spp.
Siphonaptera (Fleas)	Plague (*Pasteurella pestis*)	*Xenopsylla cheopis, X. astia, X. eridos, Nosopsyllus fasciatus, Pulex irritans, etc.*
	Endemic typhus (*Rickettsia mooseri*)	*Xenopsylla cheopis* *Nosopsyllus fasciatus*

8

PEST IDENTIFICATION

An important part of dealing with pest problems is identification of the causal agent. Identification to group level (e.g. aphids) is sometimes adequate but usually it is necessary to be more precise than this. Many features of body form and structure are used in identifying insects but the nature of plant injury is equally important in practice as this is often noticed before the presence of pests is suspected. The type of plant injury can provide useful clues to pest identification. Let us first examine the main features of major classes of arthropods.

MAIN FEATURES OF MAJOR CLASSES OF ARTHROPODS

Crustacea (crayfish, crabs, slaters, etc.)

Head and thorax merged. Two pairs of antennae. (One pair may be very poorly developed in some groups, e.g. slaters). Appendages on nearly all segments, modified as legs, swimming organs etc. Almost always aquatic. Respiration by gills or through the general body surface.

Myriapoda (centipedes, millipedes, symphilids)

No separation between thorax and abdomen. Many pairs of legs. (One or two pairs to every segment). One pair of antennae. Tracheal system present. Terrestrial.

Arachnida (spiders, scorpions, harvestmen, mites, ticks)

Head and thorax merged. (In mites all body regions are merged.) Four pairs of legs. No antennae. Tracheal system usually present. Mostly terrestrial but some aquatic forms.

Insecta (insects)

Body divided into head, thorax and abdomen. Three pairs of true (jointed) legs though additional false legs may be present. Single pair of antennae. Tracheal system present (except in lower forms). Mostly terrestrial but some groups common in fresh water.

PEST DAMAGE TO PLANTS

In practical field situations it is often just as important, if not more important, to be able to recognise plant damage caused by a specific pest, as to be able to recognise the pest itself.

Symptoms of plant injury are often the first thing that is noticed in a field situation and moreover specimens of the pest itself may be difficult to find if, for example, it is nocturnal in habit and is not found in the vicinity of damage during the day time.

Unfortunately it is impractical to provide more than a very limited amount of pest damaged fresh plant material in laboratory classes, so it is particularly important to take good note of photographs, illustrations, and descriptions of plant injury provided for specific pests. Growers of specific crops should be able to recognise all the key and occasional pests (and diseases) of each crop. The number is often quite small so that identification of the causes of pest damage is not usually too difficult.

When attempting to identify a plant pest problem consideration of the following aspects may assist greatly:

What species of plant is being attacked?	As most plant pests are rather selective this helps to narrow down the range of possibilities considerably.
What part of the plant is injured?	Individual pest species do not attack all parts of plants indiscriminately, e.g. corn earworms feed on flowers and developing fruits and rarely attack foliage.
What is the nature of plant injury?	Distinguish between *signs* (remains or byproducts of the pest) and *symptoms* (effects on the plant or plant reaction to injury). Typical insect signs are: • shed skins (aphids) • frass (excreta), (caterpillars) • honeydew (sap sucking insects) • webbing (some caterpillars) Symptoms are many and varied but can be divided broadly into two categories depending on the pest's mouth part type: *Biting and chewing* — holes eaten in leaves or other parts of the plant. Such injury may be internal, within shoots, stems or even leaves (leaf mining). Injury caused by insects with rasping mouth parts tends to be similar but the wounds formed are often very ragged. *Piercing and sucking* — wilting, stunting, distortion and bleaching but without loss of solid plant tissue. Symptoms of plant injury for particular pests are sometimes so characteristic that they can be reliably identified from these alone.
Are insects present?	If so, are they larvae (or nymphs) or adults, or even other life

	stages (eggs, pupae)? Are they closely associated with the plant injury?
	Are other organisms also present? Remember that many insects are often found on plants which are not pests and which may even be beneficial.
What time of year?	Some pests are seasonal in occurrence and only cause plant injury at certain times of the year, e.g. damage from grass grub *beetles* takes place only in spring.
What part of the country?	Several pests are limited in geographical distribution and so can possibly be eliminated from consideration.

Obviously the use of such information in helping to identify a pest problem is only as good as your own knowledge of the points concerned and how they relate to specific pest species.

One further point to emphasise is that injury to plants and associated symptoms can be caused by many other factors besides insects and other pests. These include plant disease organisms, physical factors such as sun scald, frost and wind injury and contamination with herbicides. In attempting to determine the cause of plant damage in specific instances such factors may well need to be considered if only then to be set aside as possible causes.

KEY TO INSECT ORDERS

A pest can be identified conclusively only with the help of *key*. The following pages consist of two reference keys to insect orders working with adults, larvae or nymphs. We will use this key to identify a number of nine or ten insect specimens to their order. Record the result against the number of each specimen. Read the introduction to the key carefully before you start. To reach the correct answer it is essential to work slowly and carefully and only to move on to the next step when you are certain that you have the correct interpretation of the previous one. Any key does not attempt to show any relationship between one order and another. It merely uses the method of referring the student onto further options where necessary, until the order of the insect has been established.

Do not expect the key to work 100 per cent of the time. The large insect groups are the most difficult to define, because they include such a range of variation. However, these are the groups that one can most easily learn to recognise on sight.

KEY 1

The following key (after R. Imes, 1992), which I call beginner's key, will give you a starting point from which to identify any unfamiliar insect by placing it in correct order. Begin with number one and decide which of the two options, a or b, fits the insect in question. Whichever one you choose will either reveal the order or direct you to another part of the key, and so on, until the order is disclosed. Some orders have members that are both winged and wingless, and hence are listed twice in the key.

1. Does the insect have wings?
   - *(a)* Yes — Go to # 2
   - *(b)* No — Go to # 17

2. How many pairs of wings does the insect have?
   - *(a)* One — Diptera
   - *(b)* Two — Go to # 3

3. Do the two pairs of wings differ greatly in structure, the first pair being thick and hard or leathery?
   - *(a)* Yes — Go to # 4
   - *(b)* No — Go to # 7

4. Is the first pair of wings rigid, and do they meet in a straight line down the middle of the back?
   - *(a)* Yes — Go to # 5
   - *(b)* No — Go to # 6

5. Is there a pair of prominent pincerlike cerci at the tip of the abdomen?
   - (a) Yes — Dermaptera
   - *(b)* No — Coleoptera

6. Does the insect have:
   - *(a)* chewing mouth parts, front wings leathery and heavily veined, and hind wings folded like a fan? — Orthoptera
   - *(b)* sucking mouth parts and front wings leathery at the base, membranous and overlapping at the tip? — Hemiptera

7. Are the mouth parts a coiled tube and the wings covered with scales?
   - *(a)* Yes — Lepidoptera
   - *(b)* No — Go to # 8

8. Are the wings rooflike, sloping downward and outward from the middle of the back?
   - *(a)* Yes — Homoptera
   - *(b)* No — Go to # 9

9. Is the insect slender and mothlike, with long, slender antennae and wings that are widest past the middle?
   - *(a)* Yes — Trichoptera
   - *(b)* No — Go to # 10

10. Do the wings have few or no cross veins?
    - *(a)* Yes — Go to # 11
    - *(b)* No — Go to # 12

11. Does the insect have chewing mouth parts and hind wings somewhat smaller than the front wings?
    - *(a)* Yes — Hymenoptera
    - *(b)* No — Thysanoptera

12. Are there two or three long, slender, tail-like appendages on the tip of the abdomen?
    - *(a)* Yes — Ephemeroptera
    - *(b)* No — Go to # 13

13. Does the head have an elongated trunk like beak with chewing mouth parts at its tip?
    - *(a)* Yes — Mecoptera
    - *(b)* No — Go to # 14

14. Does the insect have inconspicuous antennae, long narrow wings, and a long slender abdomen?
    - *(a)* Yes — Odonata
    - *(b)* No — Go to # 15

15. Does the insect have two short cerci on the tip of its abdomen and front wings narrower than the rear wings?
    - *(a)* Yes — Plecoptera
    - *(b)* No — Go to # 16

16. Do the tarsi each have 5 segments?
    - *(a)* Yes — Neuroptera
    - *(b)* No — Isoptera

17. Is the insect antlike, with a narrow waist?
    - *(a)* Yes — Hymenoptera
    - *(b)* No — Go to # 18

18. Is the insect antlike, but with a wide waist?
    - *(a)* Yes — Isoptera
    - *(b)* No — Go to # 19

19. Is the insect small and flattened, with chewing mouth parts and a head about as wide as its body?
    - *(a)* Yes — Go to # 20
    - *(b)* No — Go to # 21

20. Are the antennae long, and composed of many segments?
    - *(a)* Yes — Psocoptera
    - *(b)* No — Mallophaga

21. Is the insect's body soft and rounded, with two short tubes protruding from the abdomen, and with a small head?
    - *(a)* Yes — Homoptera
    - *(b)* No — Go to # 22

22. Is the insect very small, with a vertically flattened body, a hooklike claw on each leg, and sucking mouth parts?
    - *(a)* Yes — Anoplura
    - *(b)* No — Go to # 23

23. Is the insect very small and narrow (flattened laterally) with sucking mouth parts?
(a) Yes Siphonaptera
(b) No Go to # 24

24. Is the insect:
(a) delicate with chewing mouth parts and threadlike "tails" and antennae? Thysanura
(b) very small with a springlike lever folded under its abdomen which it uses for leaping? Collembola

KEY 2

The following key is adapted by A. Youdeowei (1977) from a very exhaustive key of A.D. Imms, *A General Textbook of Entomology* (1960). For the graduate and post-graduate students of entomology, this key is sufficiently informative. However, for more exhaustive key, refer to the Imms' key.

1. *(a)* Wings present 2
(b) Wingless, or with vestigial or rudimentary wings not suitable for flight 23

2. *(a)* Insects with one pair of membranous fore wings; the hind wings are modified into knobbed halteres. Mouth parts forming a proboscis, only exceptionally vestigial. Mesothorax large and fused with the pro- and metathorax. Abdomen without caudal filaments **Diptera**
(b) Insects with two pairs of wings 3

3. *(a)* Fore wings leathery, horny or parchment-like, at least at the base (called tegmina or hemielytra); differing materially in texture from the membranous hind wings which exceptionally may be absent. Prothorax large, not fused with mesothorax 4
(b) Fore wings not leathery or horny 12

4. *(a)* Fore wings containing veins, or the hind wings not folded crosswise when at rest 5
(b) Fore wings (called elytra) without veins, of uniform and horny consistency; the hind wings folded crosswise as well as lengthwise when at rest and hidden beneath the elytra. Mouth parts adapted for chewing 11

5. *(a)* Mouth parts forming a long, jointed proboscis, adapted for piercing and sucking 6
(b) Mouth parts with mandibles fitted for chewing and moving laterally 7

6. *(a)* Head opisthorhynchous; gula well-developed. Fore wings usually leathery at the base and membranous at the apical portion. The membranous parts usually overlap one another and lie flat on the abdomen when at rest **Hemiptera-Heteroptera**
(b) Head opisthorhynchous; gula absent or represented by a small membrane. Fore wings held in a triangular roof-like manner when at rest **Hemiptera-Homoptera**

7. *(a)* Hind wings similar to fore wings, and folded under the fore wings when at rest. Thickened basal part of wing very short, separated from the rest of the wing by a perforated transverse suture. Social species living in colonies **Isoptera**

(*b*) Hind wings folding, fan-like, and broader than fore wings **8**

8. (*a*) Antennae long and setaceous. Prothorax large and free from the mesothorax; cerci present; fore wings usually long. Usually large, or moderately large, species **9**

(*b*) Antennae short, flabellate, with four to seven segments, at least one joint bearing a long lateral process; cerci absent; fore wings reduced to club-shaped structures like halteres of Diptera. Pro- and mesothorax greatly reduced; metathorax large, occupying at least half of the body. Very small, active species. Short-lived insects, parasitic on other insects, especially on bees and wasps **Strepsiptera**

9. (*a*) Hind femora same size as fore femora. Body may be slender, elongate and cylindrical, or depressed and leaf-like. The wings superimposed when at rest. Terga and sterna sub-equal **10**

(*b*) Hind femora almost always much larger than the fore femora. Hind legs large and adapted for jumping; otherwise fore legs broadened and adapted for digging—Gryllotalpidae (mole crickets). Species usually capable of chirping or making a creaking noise. Body more or less cylindrical; wing held sloping against the body when at rest. Mouth parts adapted for biting and chewing. Head hypognathous **Orthoptera**

10. (*a*) Body elongate and oval; head partly concealed from above by the prothorax. Legs similar and adapted for running; or fore legs raptorial, with spines, and adapted for catching prey. Coxae large, and tibia noticeably spiny or with many bristles **Dictyoptera**

(*b*) Prothorax short; legs similar, and adapted for walking. Body may be slender, elongate and cylindrical, or depressed and leaf-like. Head prognathous; mouth parts adapted for biting and chewing **Phasmatodea**

11. (*a*) Abdomen terminated by movable, almost always heavily sclerotized, forceps (cerci). Antennae long and filiform. Fore wings short, exposing most of the abdomen; hind wings nearly circular, delicate, membranous, radially folded from near the centre. Resemble beetles **Dermaptera**

(*b*) Abdomen not terminated by forceps. Antennae of various forms. Fore wings usually covering abdomen, and modified into horny or leathery elytra which almost always meet mid-dorsally to form a straight mid-dorsal line; hind wings membranous, and folded under the fore wings when at rest. Generally body strongly sclerotized **Coleoptera**

12. (*a*) Wings long and narrow, with marginal setae, almost veinless. Tarsi one- or two-jointed, with swollen tip. Mouth parts asymmetrical, without mandibles, and adapted for piercing and sucking plant tissues; cerci absent. Generally minute insect **Thysanoptera**

(*b*) Wings broad, usually supplied with veins, but without marginal setae; rarely somewhat linear. Tarsi have more than two joints, and last tarsal joint not swollen **13**

13. (*a*) Wings, legs and body covered with scales, intermixed with hairs, which nearly always form a colour pattern on the wings. Mouth parts made of maxillae and adapted for sucking; long and coiled under the head when not in use; mandibles vestigial or absent (except micropterygoidea) **Lepidoptera**

(*b*) Wings, legs and body not covered with scales, although sometimes hairy with a few scales intermixed; sometimes with bristles, especially on the legs, or rarely with waxen flakes or dust. Colour pattern when present extending to the wing membrane **14**

14. (*a*) Hind wings with anal area separated, folded fan-wise in plaits when at rest; nearly always wider and larger than fore wings; wing veins numerous. Antennae prominent. Nymphs or larvae aquatic **15**

(*b*) Hind wings without a separated anal area, not folded and not larger than the fore wings **17**

15. (*a*) Tarsi five-jointed; cerci not pronounced **16**

(*b*) Tarsi three-jointed; cerci long with many segments; segments of the thorax all about the same size and articulated with each other; no ovipositor **Plecoptera**

16. (*a*) Wings with a number of subcostal cross veins, with both fore and hind wings similar and held in a roof-like manner over the abdomen when the insect is at rest. Wing venation primitive **Neuroptera**

(*b*) Wings without subcostal cross veins; with surfaces hairy; fore wings are elongate and narrower than the hind wings. Moth-like insects with mouth parts adapted for licking fluid **Trichoptera**

17. (*a*) Antennae long, filiform, with many segments. Two pairs of wings similar and nearly equal in size; may be spotted or banded; venation primitive. Head produced into a long rostrum with biting mouth parts at the end. Male genitalia usually greatly swollen, forming flexed bulb **Mecoptera**

(*b*) Antennae not long, not filiform. Wings of unequal size. Head not drawn out into a long rostrum. Male abdomen not forcipate **18**

18. (*a*) Wings net-veined, i.e. with many cross veins and long veins making a large number of cells **19**

(*b*) Wing venation usually reduced; few cross veins and few cells **20**

19. (*a*) Hind wings much smaller than fore wings. Abdomen ending in long thread-like processes. Tarsi four-or five-jointed. Weak fliers. Mouth parts vestigial **Ephemeroptera**

(*b*) Hind wings nearly like fore wings; both elongate, may be equal or sub-equal; venation with reticulate pattern with dark stigma towards the tip. Eyes large and conspicuous. Abdominal sterna with male genital armature. Strong and active fliers **Odonata**

20. (*a*) Wings membranous; hind wings smaller than the fore wings; fore wings and hind wings coupled by means of hooklets; wings with relatively few angular cells; costal cell without cross veins. Prothorax small and fused with the mesothorax as the

propodeum. Tarsi five-jointed, rarely three- or four-jointed. Abdomen ending in a sting or specializd ovipositor; abdomen without cerci **Hymenoptera**

(*b*) Fore and hind wings not coupled. Tarsi two- or three-jointed. Thoracic segments not fused. Abdomen not ending in a sting; abdomen with one pair of cerci, generally two-segmented 21

21. (*a*) Wings with smoky appearance, with narrow clear areas between the principal veins. Tarsi three-segmented; first segment of tarsi of the prothoracic legs are swollen and bear silk glands; cerci two-segmented and are conspicuous and asymmetrical in males; fragile soft-skinned insects **Embioptera**

(*b*) Wings not smoky in appearance 22

22. (*a*) Wings delicate, venation prominent but reduced, radial sector and media branched. Tarsi two- or three-segmented; cerci absent. Head very mobile; compound eyes markedly convex, protruding from surface of head **Psocoptera**

(*b*) Wings deciduous like those of the Isoptera; venation greatly reduced; radial sector and media simple unbranched. Tarsi two-segmented; cerci short, one-segmented **Zoraptera**

23. (*a*) Body more or less insect-like, with distinct head, thorax and abdomen, jointed legs, and capable of locomotion 24

(*b*) Body not clearly divided into head, thorax, and abdomen 28

24. (*a*) Terrestrial, breathing through spiracles rarely, without special respiratory organs; free living 25

(*b*) Parasitic on warm-blooded animals 30

25. (*a*) Mouth parts retracted into the head and scarcely visible or not at all visible. Underside of abdomen with stylets or other pregenital appendages. Delicate and minute insects **Apterygota (Subclass)**

(*b*) Mouth parts conspicuously visible externally. Antennae always present. Underside of abdomen rarely with stylets 26

26. (*a*) Antennae absent. Mouth parts adapted for piercing. Thorax with clearly defined segments; prothorax reduced. Abdomen with eleven segments, and the first three abdominal segments carry a pair of appendages. No long cerci. Tarsi one-segmented **Protura**

(*b*) Antennae conspicuous. Long cerci, or basal ventral tube, present on the abdomen 27

27. (*a*) Abdomen consisting of six segments or less, with forked sucker (collophore) at base of abdomen. No long cerci. Usually with conspicuous springing organ (furcular) near end of abdomen. Mouth parts adapted for biting. No compound eyes. Antennae with four segments **Collembola**

(*b*) Abdomen consisting of more than eight visible segments; with long and multi-articulate cerci or strong sclerotized forceps at the end. Compound eyes and ocelli absent **Diplura**

28. (*a*) Mouth parts mandibulate, adapted for chewing **29**
(*b*) Mouth parts adapted for piercing and sucking **30**

29. (*a*) Body usually covered with scales. Abdomen with three prominent caudal filaments, and bearing at least two pairs of ventral styles. Tarsi with two or four segments **Thysanura**
(*b*) Body not covered with scales. Abdomen without caudal filaments **31**

30. (*a*) Body laterally compressed. Compound eyes absent, but two ocelli may be present. Antennae three-jointed and stout, lying in grooves in the head. Coxae of legs very large, and tarsi are five-jointed. Cerci and ovipositor absent. Jumping species **Siphonaptera**
(*b*) Body dorso-ventrally flattened. Compound eyes very small or absent; no ocelli. Pro-, meso-, and metathoracic segments fused. Tarsi one-segmented, ending in a claw. No cerci **Siphunculata**

31. Mouth anterior. Generally elongate-oval insects, with somewhat triangular head. Tarsi with one or two segments, ending in a single or sometimes paired claw. Parasites of birds or mammals **Mallophaga**

9

INSECTICIDE APPLICATION EQUIPMENT

Efficiency of pest control by insecticides is dependent on the correct type of pesticide formulation applied at the proper time and correct dosage, besides the good equipment. Equipment used to apply insecticides must be suited to the crop (or other situation in which the pest is to be controlled), to the scale of operation, and to the formulation of chemical to be applied. In the long run, it is good economy to buy the best equipment available. Each piece of equipment should be chosen so it will be large enough to do the required job but not so large that it will be unnecessarily cumbersome. Stainless steel tanks are preferable to those made of other materials, valves should be easily operated and should provide a quick and complete shut off; hoses should be the smallest and lightest size possible to do the job adequately; nozzle tips should discharge the chemical in proper amounts and in the best manner.

Application equipment of high quality means not only good, durable construction but also good design that gives it a professional appearance. Because it is desirable and often necessary to utilize pesticides in a variety of ways, various types of dispensing equipment have been developed. This equipment is placed in the following categories: (1) Sprayers, (2) Dusters, and (3) Granule Applicators. In addition, specialised equipment of various sorts (such as mist machines and fog generators) are available for particular purposes. Since detailed consideration of pesticide application equipment is out of the scope of this book, only a brief consideration is given here to some important features of the main types of equipment.

1. Sprayers

Sprayers vary greatly in size from the hand-pumped flit pump with a tank capacity of as little as 1 cup to large hydraulic machines powered by gasoline engines and having tanks which can hold several hundred litres of pesticide formulation. All sprayers must provide some means of breaking up the spray liquid into droplets (*atomising*) and of conveying the droplets onto the object to be sprayed. There is always a tank, a method of pressurizing the liquid, a delivery line leading to a valve, and another delivery line leading from the valve to a nozzle (Fig. 9.1). All other items found on any sprayer such as agitators, pressure control valves, pulse dampers, pressure releases, and others are merely additon to the basic sprayer and modify the performance of the basic parts in some manner.

In most smaller machines hydraulic pressure provides the means both of atomising the spray liquid and of imparting momentum to the droplets so that they carry to and impinge on the target. Many larger machines, however, especially those designed for low volume application, employ a fan generated stream of air to carry the droplets. They are referred to as *air assisted* or *air blast sprayers*. Atomisation in such sprayers may be through normal hydraulic nozzles or by power operated spinning discs or cages, which provide more uniform droplet size than conventional nozzles.

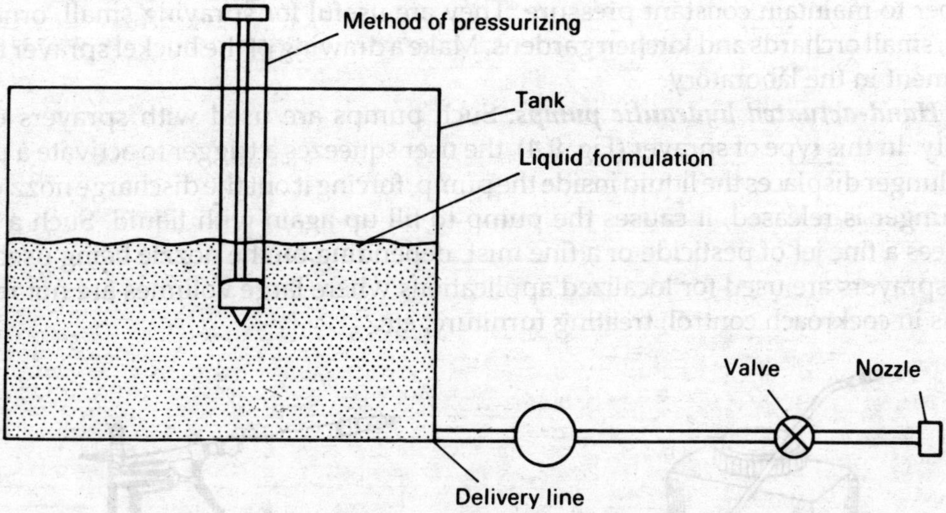

Fig. 9.1: Components of a basic sprayer.

(*i*) ***Compressed air sprayers:*** A compressed air sprayer is a small, hand-operated, hand-carried spray equipment with a capacity in the range of 6 to 15 litres. Air pressure is supplied by a hand-operated pump and is contained in the tank above the surface of the liquid to be dispensed. This air, usually compressed to 20-50 pounds pressure, forces the pesticide out through a discharge tube to the nozzle when the discharge valve is opened (Fig. 9.2). The liquid is not mixed with air but is rushed out as a wet spray without atomization. Some of these sprayers are equipped with a pressure indicating gauge. Compressed air sprayers are used for insects in lawns, orchards, home and in poultry houses. They are used by public health workers for applying residual deposits of insecticides in homes and other buildings to control mosquitoes and other insects of medical importance.

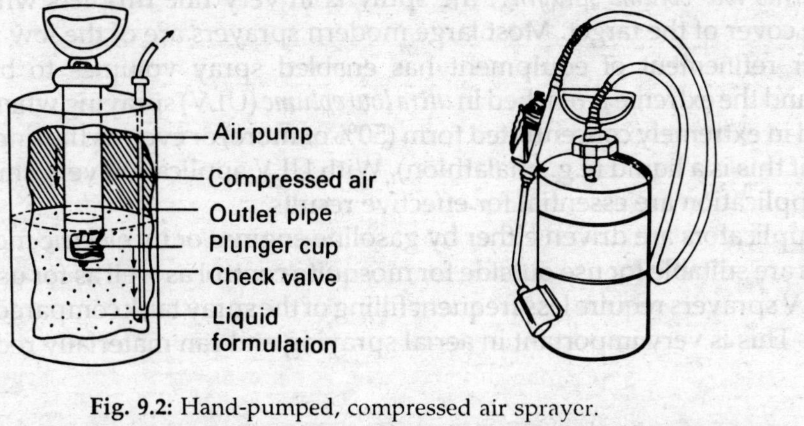

Fig. 9.2: Hand-pumped, compressed air sprayer.
Large arrow shows of air presence in liquid, and small arrows show direction of liquid flow.

(*ii*) ***Knapsack sprayers:*** A knapsack sprayer is carried on the back of the operator and the pressure is created by the user working a lever up and down (Fig. 9.3). It has a tank capacity of 8 to 20 litres and has higher pressures generated than a compressed air sprayer, so it is better suited for large gardens.

(*iii*) ***Bucket-type sprayers:*** These sprayers are fitted with plunger type pumps and an air chamber to maintain constant pressure. They are useful for spraying small, ornamental plants, small orchards and kitchen gardens. Make a drawing of the bucket sprayer from the equipment in the laboratory.

(*iv*) ***Hand-actuated hydraulic pumps:*** Such pumps are used with sprayers of small capacity. In this type of sprayer (Fig. 9.4), the user squeezes a trigger to activate a plunger. This plunger displaces the liquid inside the pump, forcing it out the discharge nozzle. When the plunger is released, it causes the pump to fill up again with liquid. Such a sprayer produces a fine jet of pesticide or a fine mist, depending on the nozzle being used. These small sprayers are used for localized applications where large volumes are not required, such as in cockroach control, treating furniture, etc.

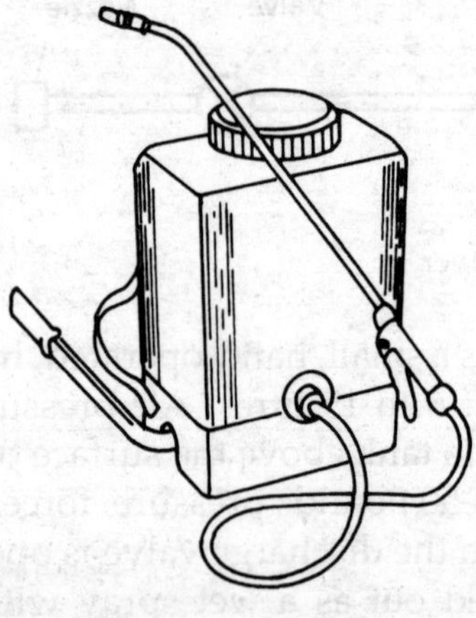

Fig. 9.3: Knapsack sprayer

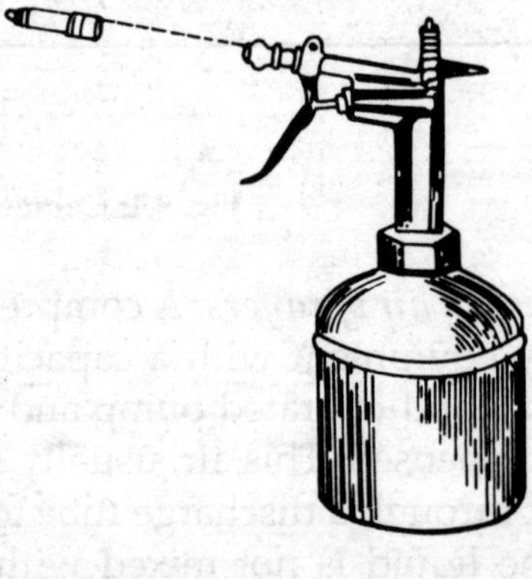

Fig. 9.4: Hand-actuated hydraulic pump sprayer.

(*v*) ***ULV devices:*** Previously all sprayers were of the hydraulic type and applied pesticides in very dilute form, almost wetting completely the target surfaces. But the development of air blast machines has permitted the use of chemicals in more concentrated form. In this *low volume spraying*, the spray is in very fine droplets which do not give complete cover of the target. Most large modern sprayers are of the low volume type.

Further refinement of equipment has enabled spray volumes to be progressively reduced and the extreme is reached in *ultra low volume* (ULV) spraying where the insecticide is applied in extremely concentrated form (50% or more) or even as the undiluted technical material if this is a liquid (e.g., malathion). With ULV application very small droplets and precise application are essential for effective results.

ULV applicators are driven either by gasoline engines or by electric motors. The larger machines are suitable for use outside for mosquito control as well as for use inside in large areas. ULV sprayers require less frequent filling of the spray tank compared to high volume spraying. This is very important in aerial spraying and can materially reduce application costs. Also, compared to high volume spraying, less pesticide is normally used to achieve the same result as more is deposited on target.

Sprayer Accessories

As mentioned earlier, the basic parts of any sprayer are the same. There are, however, many accessory parts which are added to sprayers to modify or control the action of the basic parts (Fig. 9.5): power source, pump, agitators, pressure regulators, spray tanks, strainers, pressure gauges, nozzles, valves, tank shut-off valves, and hose.

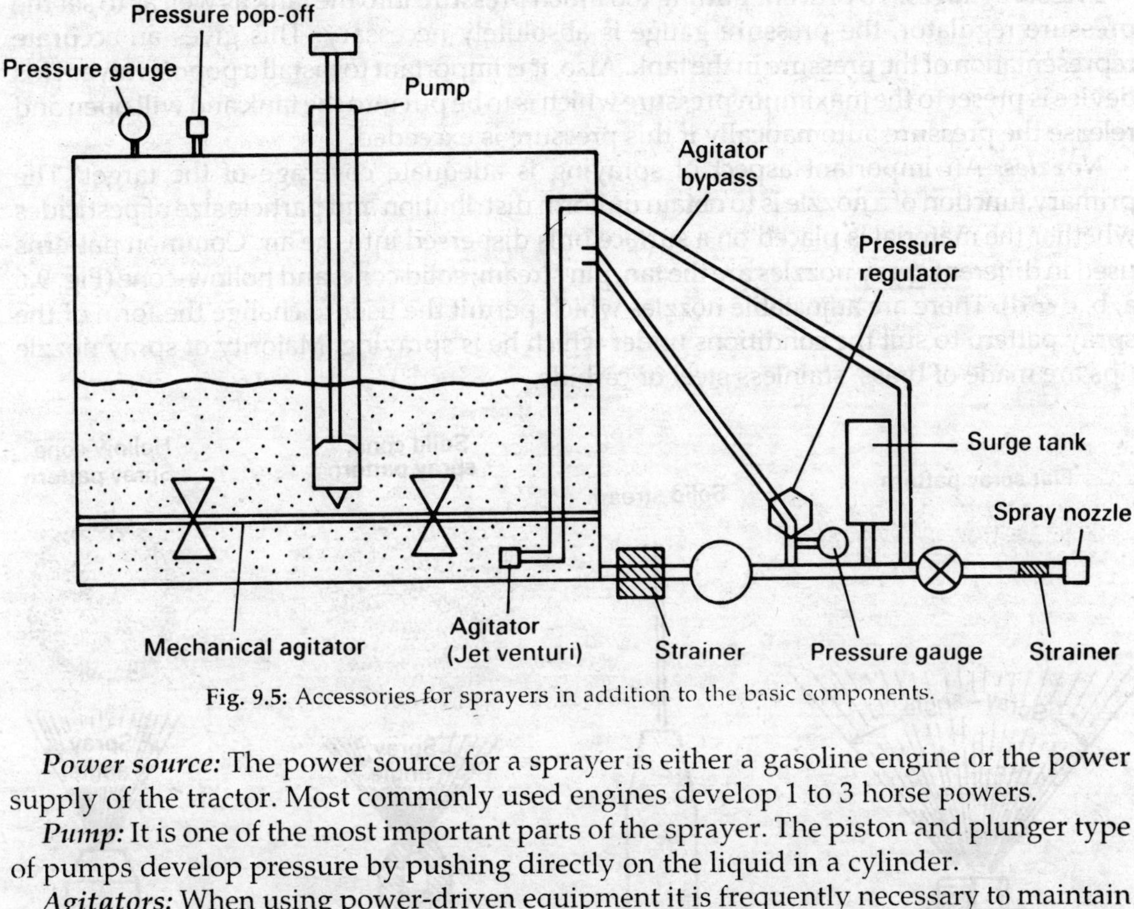

Fig. 9.5: Accessories for sprayers in addition to the basic components.

Power source: The power source for a sprayer is either a gasoline engine or the power supply of the tractor. Most commonly used engines develop 1 to 3 horse powers.

Pump: It is one of the most important parts of the sprayer. The piston and plunger type of pumps develop pressure by pushing directly on the liquid in a cylinder.

Agitators: When using power-driven equipment it is frequently necessary to maintain an active agitation of the liquid in the tank in order to keep it thoroughly and evenly mixed. If emulsion type materials are being used, a simple return by-pass as found on most power sprayers is sufficient to keep the emulsion in suspension. Mechanical agitators provide excellent mixing of the chemical for wettable powder solutions. Just be sure that your pump has adequate capacity to supply both the agitator and the nozzle.

Pressure regulators: In all power pump operated equipment it is necessary to have a pressure regulator located in the line between the pump and the discharge valve to the nozzle. The pressure regulator is set at the desired pressure for spraying and, when the pressure in the line exceeds this pressure, the regulator opens. This permits some of the chemical to return to the tank through a by-pass line thus lowering the pressure at this point.

Spray tanks: Spray tanks can be made of wood, steel, stainless steel, glass, fibreglass, and plastic. However, tanks should be strong and durable. The most suitable tanks for use in everyday pest control operations are stainless steel and fibreglass.

Strainers: A strainer is necessary in all sprayers to prevent damage to pumps and to prevent clogging of nozzles. In large equipment a strainer is placed between the tank and the pump to prevent coarse particles from reaching the pump and damaging it. Also, another strainer is frequently placed in the nozzle itself to hold back any particles liable to block the nozzle tip.

Pressure gauges: To prevent putting too much pressure into the tank as well as to set the pressure regulator, the pressure gauge is absolutely necessary. This gives an accurate representation of the pressure in the tank. Also, it is important to install a popoff valve. This device is preset to the maximum pressure which is to be put into the tank and will open and release the pressure automatically if this pressure is exceeded.

Nozzles: An important aspect of spraying is adequate coverage of the target. The primary function of a nozzle is to obtain uniform distribution and particle size of pesticides whether the material is placed on a surface or is dispersed into the air. Common patterns used in different spray nozzles are the fan, pin stream, solid cone, and hollow-cone (Fig. 9.6 a, b, c & d). There are adjustable nozzles which permit the user to change the form of the spray pattern to suit the conditions under which he is spraying. Majority of spray nozzle tips are made of brass, stainless steel or carbide.

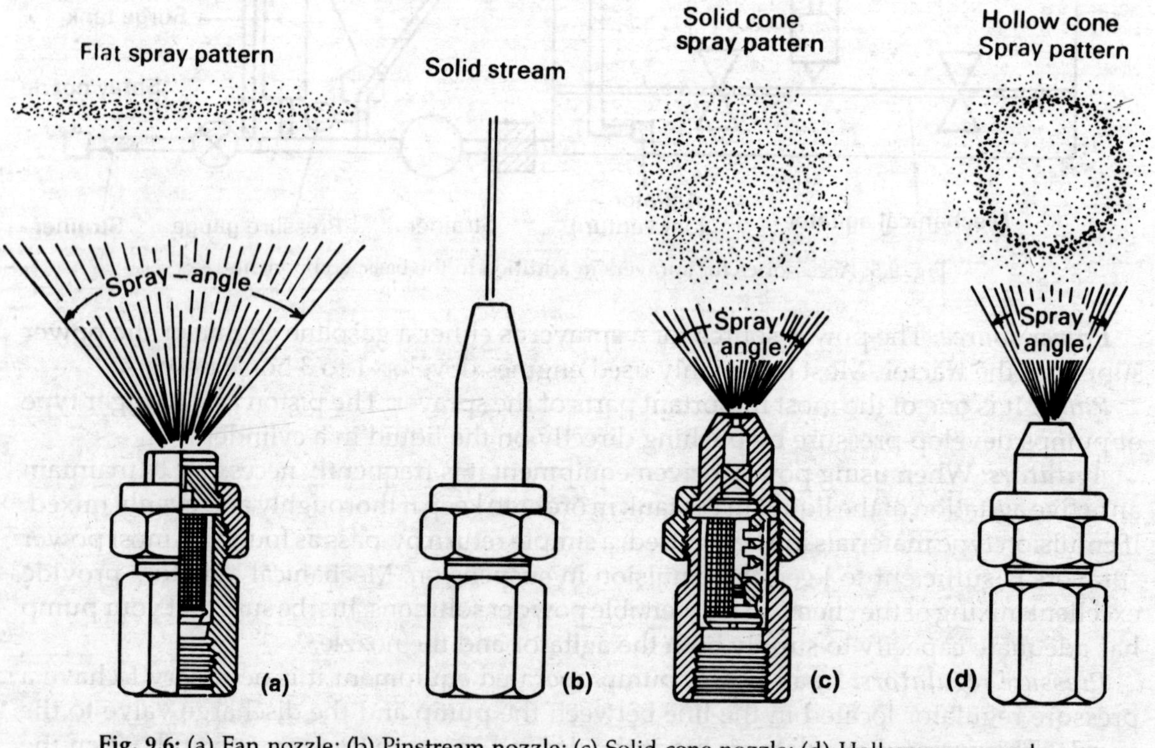

Fig. 9.6: (a) Fan nozzle; (b) Pinstream nozzle; (c) Solid cone nozzle; (d) Hollow cone nozzle

Valves: Valves provide a means of turning on and shutting off the flow of material to the nozzle in a continuous flow sprayer. Some valves are turned, some are squeezed while some are bent to cause them to operate.

Tank shut-off valves: Such valves should be provided between the tank and the hose, so pressure can be shut off from the hose without shutting down the power source driving the equipment.

Hose: The hose on any sprayer should be long enough for the purpose intended, should be of sufficient diameter to carry an adequate flow of chemicals and should be made of materials which will not be deteriorated by the pesticides and solvent being used. Different materials like neoprene (synthetic rubber), polyethylene nylon and tygon (plastics) are used in making the hoses. Where high pressure spraying is to be done it is necessary to use heavy duty hose which will withstand high pressure.

2. Dusters

Since dusts are used far less than sprays as a means of applying insecticides to plants, there has been less development of dusting equipment. Dusters, like sprayers, may be small or large, hand or power operated. Most dusters incorporate a metering device of some sort to regulate dust output and a blower to carry the dust to the target. Large, power operated dusters are best suited for agricultural work and some types of mosquito control. There are also available small compressed air tanks for applying dust. Air is compressed in the tanks and when dust is desired the working of a single lever causes the compressed air to discharge dust.

Make a drawing of a duster in your notebook.

3. Granule Applicators

Various types of granule application equipment are available. Selection for a particular purpose depends on whether the granules are to be spread (e.g. over pasture) or applied in a restricted manner (e.g. in furrow application at sowing time). Most granule applicators incorporate a mechanism to feed granules in regulated amounts from the hopper. For in furrow application, granules may be simply released "down the spout" into the seed furrow. The granules so put are lightly "incorporated" into the soil by incorporation wheels, chain drags or press wheels. For broadcast application there is usually some means of throwing the granules to form a swath, e.g., rapidly rotating ridged disc or oscillating arm. Granule applicators, like sprayers, are calibrated. The amount of granules spread by applicators depends on the size of the feeder-gate setting (from which granules flow), speed of travel, field roughness, and the flow rate of granules (which is affected by size, shape, texture, and density of granules, as well as by temperature and humidity).

10

LABORATORY INSECTICIDE EXPERIMENTS

A series of laboratory experiments will be undertaken using three different insecticides (lindane, malathion and permethrin) to demonstrate several properties of insecticides using stored grain pest beetles (*Tribolium* or *Trogoderma*) and mosquitoes as test insects. The properties to be investigated include: comparative toxicity by contact, persistence and fumigant action of the compound.

1. Comparative toxicity (by contact action)

Two basic methods are available :

(*a*) exposure of insects to a treated surface or medium;

(*b*) direct treatment of individual insects with small measured droplets of solution (topical treatment).

In the former type of treatment the amount of insecticide picked up by the insect (dose) will depend on the insect's activity, the nature of the deposit, and the duration of exposure. The exact dose will be unknown, therefore we refer to *dosage* (amount per unit area/volume of medium treated) rather than dose (amount *received* by each insect). Exposure methods require less equipment and are simpler to operate. Such methods will be used throughout these laboratory exercises.

With either type of treatment, we need to run a series of concentrations of each insecticide with batches of test insects at each concentration to produce a range of effect from 0-100 per cent. Parallel control experiments are run without the test compound, in identical conditions. If the mortality is 20 per cent or more in the control, the treatment is repeated. Overall toxicity of the compound is assessed as per cent total mortality corrected for controls by Abbott's formula (1925):

$$\% \text{ corrected mortality} = \frac{\% \text{ test mortality} - \% \text{ control mortality}}{100 - \% \text{ control mortality}} \times 100$$

The corrected per cent mortality values are then plotted — effect against concentration which gives a sigmoid or bell-shaped curve. When concentration of the compound (converted into logarithmic scale) on X-axis is plotted against the toxicity (changed into probit values) on Y-axis, on the logarithmic probability paper, a straight line is drawn by best eye-fitting line method. Such a line is called **dosage-mortality-regression line** (or d-m-r line).

The LD_{50}, LD_{90} and slope of the regression line are then computed. Insecticides are then conveniently compared by means of their LD_{50} values. LD_{50} or lethal dosage$_{50}$ is that dosage of the chemical required to kill 50 per cent of the test animals. The lower the LD_{50} value the more toxic is the substance because a smaller amount is required to produce the same effect compared to a larger amount of a less toxic substance.

Procedure for Trogoderma adult and larvae

Prepare a series of graded concentrations of the three insecticides to be examined in acetone as solvent. Each class group should work with **one** insectictide each week. Take 2 or 3 replicates for each concentration.

Label a series of filter papers by marking in *pencil* to give one paper at each concentraion of insecticide. Mark with name of insecticide, concentration, and class group number. Treat each paper with 1 ml of insecticide solution from the series of dilutions provided including a solvent blank. *Do not pipette by mouth* — use the safety bulbs provided. Place the treated papers over a set of inverted nails to dry. Transfer treated papers, treated surface uppermost, to petri dishes, *handling with forceps*. Count 10 beetles or their larvae into each petri dish and cover them with the other half. Petri dishes containing beetles are incubated at 37° C or 30° C for 24 hours.

Assess your treatments 24 hours after setting up. *Method of assessment** is important for consistent results. Remove all beetles into the lid of the petri dish. Replace the filter paper into the base of the dish. Tip all beetles into the centre of the paper. Wait one minute. Count all individuals that fail to reach the margin of the dish by that time as **affected**. Record the number of affected beetles out of 10 for each dish.

A series of mortalities or effects should be obtained ranging from 0-100 per cent. Calculate the corrected mortality and plot a d-m-r line. After computing the LD_{50} and LD_{90} values, compare these in all the three insecticides provided. This will tell you the comparative toxicity of the compounds or differential susceptibility of beetles to these compounds. Discuss your results.

Procedure for Mosquito Larvae

Treatment of mosquito larvae with the chemicals is performed according to the procedure given by WHO. For the treatement, twenty- five fourth instar larvae are placed in 100 ml of dechlorinated water in small plastic bowls while 149 ml of dechlorinated water is put in 16 oz. glass jars. Stock solutions of all the compounds are prepared in acetone as follows :—

Let us assume we want a stock solution of 64 ppm (parts per million) as final concentration in the 250 ml of treatment medium.

1 ppm means		1 part in 10^6
	or	1 g in 10^6 ml
	or	1 mg in 10^3 ml
	or	1 mg in 1 litre

*An objective method of assessment of this type is essential, as varying degrees of incapacitation (leading eventually to death or to recovery) can occur from insecticidal poisoning.

To get 64 ppm dissolve 64 mg in 1 litre

or

0.064 mg in 1 ml

To get 64 ppm stock solution w.r.t. 250 times dilution, dissolve 0.064 mg × 250 = 16 mg/ml.

Stock solutions thus prepared are serially diluted with acetone to get a range of concentrations (32 ppm, 16, 8, 4, 2, 1, 0.5 .... ppm). Test concentrations are prepared by pipetting 1 ml of the appropriate dilution in each of the test jars and shaking vigorously. Mosquito larvae are added to them from the small bowls. Parallel control experiments are run by adding 1 ml of acetone alone to the test jars containing water. The test jars are kept in the insectary, under optimum conditions for 24 hours. For assessment of mortality, dead or moribund larvae are removed from the jars and counted. Overall toxicity of the insecticide is assessed after calculating the corrected mortality.

Draw a graph between concentration and per cent mortality as well as a d-m-r line. Discuss your results for all the three insecticides tested against mosquito larvae.

Procedure for Mosquito Adults

Like larval mosquitoes, adult mosquito susceptibility test described here is based on WHO procedure. The adult mosquitoes, preferably males (because females survive better and show lower mortalities) are exposed to the insecticide-impregnated filter papers in glass vials (Standard kits are also available from WHO). Filter paper of the size of the vial is cut, and suspended from the metal clamps or put on inverted nails (just to make it away from any surface). Then 1 ml of insecticide solution is put on this paper with a microsyringe. The impregnated filter paper is left to dry and final concentration of the insecticide is calculated (1 ml of compound concentration ÷ area of filter paper). These treated filter papers are used as lining on the inner side of a glass vial. Presorted 20 adult male mosquitoes are introduced in each vial and allowed to remain there for 24 hrs in the insectary (under optimum conditions of temperature and humidity). Similarly, filter paper impregnated with acetone alone serves as control. Both treatment and control should have at least 2 replicates each.

Mortality counts are made after 24 hrs. Affected specimens that are unable to walk are to be counted as dead. Construct the dosage-mortality-regression line from the results obtained in all the replicates at the chosen concentration. Plot the results on the log-probit paper. Also, record the maximum, minimum temperature of the insectary as well as relative humidity (% R. H.). Discuss the susceptibility levels of adults to various insecticides tested.

2. Persistence of the Insecticide

The persistence of an insecticide refers to the duration of toxic effect from a deposit or residue on a surface or other substrate. Some insecticides are very short lived in this respect (*non-persistent, transient*) and others persist for quite a long time (*persistent*). We will compare the persistence of a particular insecticide after different durations.

From the dosage/mortality graph from the earlier experiment, a dosage will be selected which gives approximately 95 per cent mortality (LD_{95}). Filter papers are treated with this dosage and left to age for various periods of time, say 1, 2, 4, 6 and 8 weeks. Re-assessment of mortality effects then gives a measure of persistence. At each interval, set up papers in a petri dish and expose 10 *Trogoderma* beetles. Assess after 24 hours as before. By doing this, we should be able to compare insecticides as to how quickly insecticidal activity has declined. By reference to our original dosage/mortality curves, we may be able to estimate the *amount* of insecticide remaining at each age interval. Discuss whether the insecticides provided are persistent or not.

3. Fumigant Activity of the Insecticides

Most insecticides are to some extent volatile, and if sufficiently so, may show insecticidal activity at a distance by fumigant action. To test for fumigant action, treat the filter papers at one-time or two-times of LD_{95} values of the insecticide. When dry, the filter papers are placed in the lower half of a petri dish. A second dish half *of exactly the same diameter* is then prepared—by spreading a piece of nylon net over the opening, closing with a rubber band round the rim, as in Fig. 10.1. Ten *Trogoderma* beetles are already placed in this half which

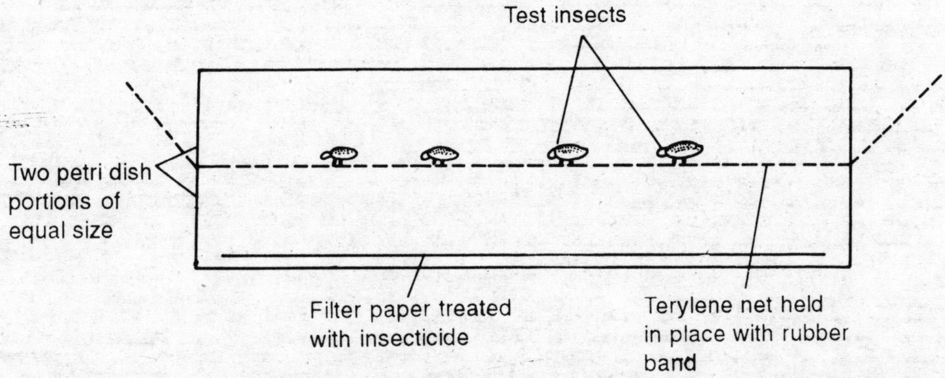

Fig. 10.1: Set up for testing for fumigant activity.

is inverted over the lower portion containing the impregnated filter paper. The reason for the two dish halves needing to be of the same size will now be apparent. Assess after 24 hours at 30^0 C as for previous experiments. Discuss your results and comment on the fumigant activity of the insecticides provided to you.

BIBLIOGRAPHY

Borror, D.J., De Long, D.M. 1971. *An Introduction to the Study of insects*. 3rd ed. Holt, Rinehart & Winston, New York, 812 pp.

Chapman, R.F. 1982. *The Insects: Structure and Function*. 3rd ed. Hodder & Stoughton, London. 919 pp.

Daly, H.V., Doyen, J.T., Ehrlich, P.R. 1978. *An Introduction to Insect Biology and Diversity*. McGraw-Hill, New York. 664 pp.

Davies, R.G. 1988. *Outlines of Entomology*. 7th ed. Chapman & Hall, New York. 408 pp.

Dennis, C.J. 1965. *Laboratory Manual for Introductory Entomology*. WMC Brown Co. Publishers, IOWA. 92 pp.

Elzinga, R.J. 1987. *Fundamentals of Entomology*, 3rd ed. Prentice-Hall, Inc. 456 pp.

Fenemore, P.G., Prakash, A. 1992. *Applied Entomology*. Wiley Eastern Ltd., New Delhi. 298 pp.

Fox, R.M., Fox, J.W. 1964. *Introduction to Comparative Entomology*. Reinhold Publishing Corp., USA. 450 pp.

Hill, D.S. 1983. *Agricultural Insect Pests of the Tropics and Their Control*, 2nd ed. Cambridge University Press, Cambridge. 750 pp.

Imes, R. 1992. *The Practical Entomologist*. Simon & Schuster Inc., New York (A Fireside Book). 160 pp.

Insects — The Yearbook of Agriculture, 1952. USDA, Washington, D.C. Oxford & IBH Publishing Co., New Delhi. 780 pp.

Oldroyd, H. 1970. *Collecting, Preserving and Studying Insects*. Hutchinson & Co. (Publishers) Ltd., London. 328 pp.

Pfadt, R.E. (ed.) 1978. *Fundamentals of Applied Entomology*. MacMillan, New York. 798 pp.

Prakash, A. 1984. *Studies on Certain Chitin Synthesis Inhibitors in Four Species of Anopheline and Culicine Mosquitoes*. University of Delhi, Ph. D. Thesis.

Rappaport, R. 1992. *Controlling Crop Pests and Diseases*. The MacMillan Press Ltd., London. 106 pp.

Richards, O.W., Davies, R.G. 1977. *Imms' General Textbook of Entomology*, 10th ed., Vol. 1. Structure, Physiology and Development. Chapman & Hall, London. 418 pp.

Richards, O.W., Davies, R.G. 1977. *Imms' General Textbook of Entomology*, 10th ed., Vol. 2. Classification and Biology. Chapman & Hall, London. 1354 pp.

Richards, O.W., Davies, R.G. 1978. *Imms' Outlines of Entomology*, 6th ed. Chapman & Hall, London. 254 pp.

Romoser, W.S. 1981. *The Science of Entomology*, 2nd ed. MacMillan, New York. 575 pp.

Ross, H.H., Ross, C.A., Ross, J.R.P. 1982. *A Textbook of Entomology*, 4th ed. John Wiley & Sons, New York. 696 pp.

Snodgrass, R.E. 1935. *Principles of Insect Morphology*. McGraw Hill. 667 pp.

Torre-Bueno, J.R. de la. 1962. *A Glossary of Entomology*. Brooklyn Entomological Society, New York. 336 pp.

Truman, L.C., Bennett, G.W., Butts, W.L. 1982. *Scientific Guide to Pest Control Operations for Pesticide Applicator Training*, 3rd ed. Harcourt Brace Jovanovich Inc. 276 pp.

Wiggelsworth, V.B. 1972. *The Principles of Insect Physiology*, 7th ed. Chapman & Hall, London, 827 pp.

Wiggelsworth, V.B. 1974. *Insect Physiology*. Chapman & Hall, London. 166 pp.

Youdeowei, A. 1977. *A Laboratory Manual of Entomology*. Oxford University Press, Nigeria. 205 pp.

GLOSSARY

Abdomen—the posterior of the three body regions of an insect.
Acari—the order of arthropods within the class Arachnida that comprises the mites and ticks.
Acaricide—(alternative term, miticide)—chemical used for control of mites.
Accessory gland—gland associated with the insect reproductive system but fulfilling some function secondary to the reproductive process. Present in both sexes of most insects.
Acetyl choline—a chemical substance produced at nerve connections (synapses) each time a nerve impulse is transmitted.
Action threshold—see economic threshold.
Active ingredient (a.i.)—the biologically active chemical in pesticide preparations.
Adulticide—a chemical to kill the adult stage.
Aedeagus—the male copulatory organ of an insect, usually taken to include accessory structures besides the penis.
Aerosol—very finely divided particles (of insecticide solution) in the air, often applied to the pressurised canisters which generate an aerosol.
Aestivate—to enter a state of dormancy over the summer period. cf. hibernate.
A.I.—sec active ingredient.
Air assisted sprayer—a sprayer in which an air stream is employed to help carry spray droplets to the object being sprayed.
Air blast sprayer—see mist blower.
Alate—possessing wings. cf. apterous.
Ametabola—(alternative term, Apterygota)—subclass of primitive wingless insects that do not undergo metamorphosis. Comprises the orders Thysanura, Diplura, Protura and Collembola. cf. Metabola.
Anal ring—an elevated ring like structure surrounding the anus in coccids.
Anoplura—an alternative name for the insect order Siphunculata which comprises the sucking lice.
Antenna (pl Antennae)—sensory feelers located on the head of insects bearing minute organs of touch and smell.
Anther—the male part of a flower which produces pollen.
Antibiosis—a mechanism of plant resistance to pests whereby harmful effects occur to the insect through feeding on the plant.
Aorta—the forepart of the dorsal blood vessel of insects. In contrast to the hindpart (heart), the arota is without valves. It discharges into the head region.
Aphaniptera—an alternative name for the insect order Siphonaptera which comprises the fleas.
Apiculture—the culture of bees, honeybees in particular.
Apodeme—an internal projection of the exoskeleton of insects and serving for muscle attachment.
Apodous—without legs.
Appendage—an external limb or similar structure arising from segments of the head, thorax or abdomen.
Apterous—without wings, cf. alate.

Apterygota—(alternative term, Ametabola)—an alternative term to Ametabola for the subclass of primitive insects that are wingless and do not undergo metamorphosis. cf. Pterygota.

Arachnida—a class of arthropods closely related to insects which includes spiders, mites, tick and scorpions.

Araneae—the order of arthropods within the class Arachnida that comprises the spiders.

Arthropleona—a sub-order of springtails (Collembola) having a distinctly segmented body of elongate form, cf. Symphypleona.

Arthropoda—the phylum of animals to which insects and other groups belong and characterised by the possession of an external jointed skeleton, segmented body and jointed limbs.

Atomise—to break up into fine droplets (as applied to spray liquids).

Attractant—a chemical or physical source which induces insects to move towards it. cf. repellent.

Augmentation—mass culture and local release of parasites, predators or pathogens to provide short term biological control of pest species.

Bait—an insecticide formulation to which a pest is attracted and which is usually ingested by the pest.

Behavioural resistance—a mechanism in insects of resistance to insecticides which involves a bahavioural change whereby contact with the insecticide is avoided.

Binomial system—the system of dual latin names for plants and animals devised by the Swedish naturalist Linnaeus and now almost universally applied to living organisms.

Biological control—the use of living organisms to control pest species.

Biomass—the weight of living organisms in any given habitat.

Bionomics—the study of the habits, life histories and adaptations of living organisms.

Biotic potential—(alternative term, reproductive potential)—the maximum possible rate of increase of an organism in the absence of any limiting factors.

Biotype—a genetically distinct strain or sub-group of a species distinguished by some behavioural or physiological difference but indistinguishable morphologically. Usually applied to strains of insect species which can overcome resistant plants.

Boll—the pod in which the lint and seeds of cotton develop.

Brachypterous—having short underdeveloped wings that do not cover the abdomen and are generally incapable of flight.

Broad spectrum—(as applied to insecticides)—having an effect on a wide range of insects.

Brood—a group of individuals of a given species in insects that have hatched into larvae or have become adult at approximately the same time, and live together in a defined and limited area.

Bubonic plague—a bacterial disease of rodents and man caused by *Pasteurella pestis* and transmitted chiefly by the oriental rat flea.

Bursa copulatrix—a specialised part of the female insect reproductive system into which sperm is received from the male during copulation.

Campodeiform—a form of insect larva (or hymph) with well developed legs, prominent antennae and usually several tail filaments.

Capitulum—a small head; in particular the false head of certain mites and ticks on which the mouthparts are borne.

Carbamate (insecticide)—a group name applied to insecticides which are based on the carbamic acid nucleus. Carbaryl is an example.

Carcinogenic—cancer inducing.

Cardo—a basal sclerite forming part of the maxilla in the mouthparts of insects.

Caste—structurally and functionally distinct group within the colony of a social insect, eg. worker caste.

Caterpillar—a polypod or eruciform larva of a moth, butterfly, or sawfly.

Cellular rearing—rearing of individual eggs.

Cercus (pl. cerci)—one of pair of appendages at the tip of the abdomen of insects.

Chelicerae—the inner portions of the mouthparts of mites; modified to piercing stylets in plant feeding mites.

Chemical control—the use of chemicals to kill, deter or in any way suppress pest populations.

Chemoreception—sensory perception of chemical substances; includes both taste and smell.

Chemosterilant—a chemical which has the property of sterilising an insect but which does not kill it.

Chilopoda—the order of arthropods within the class Myriapoda that comprises the centipedes.

Chitin—a major constituent of the cuticle of insects; chemically a nitrogenous polysaccharide.

Chlorinated hydrocarbon—see organochlorine.

Chlorosis—a yellowing of the foliage of plants due to lack of chlorophyll.

Choline esterase—an enzyme which breaks down acetyl choline. The latter is formed at nerve connections (synapses) each time a nerve impulse is transmitted.

Chronic toxicity—the toxicity of a pesticide (to higher animals) when administered in small sub-lethal doses over a long period of time.

Chrysalis—the pupal stage of butterflies and moths.

Cibarial pump—see cibarium.

Cibarium—(alternative term, cibarial pump)—a suctorial pump present in the head region of insects with piercing and sucking mouthparts (Hemiptera).

Claspers—a pair of processes at the end of the abdomen of male insects which serve to clasp the female during copulation.

Class—a division of the animal kingdom below a phylum but above an order, eg, the class Insecta.

Coarctate—pupal form in which the last larval skin in retained as an extra covering; occurs particularly in higher Diptera.

Cocoon—a silken covering produced by some insects which enclosed and protects the pupa.

Coleoptera—the order of insects comprising the beetles (including weevils).

Collembola—the order of insects comprising the springtails.

Compatible (as applied to pesticides)—safe to mix together.

Compatibility (as applied to pesticides)—refers to whether or not different pesticides can be mixed together for application. Safe mixtures are compatible. Unsafe mixture are incompatible.

Complete metamorphosis—metamorphosis in which the insect proceeds through four distinct development stages—egg, larva, pupa and adult. cf. incomplete metamorphosis.

Contact insecticide—one which is able to kill insects by contact with the external body surface.

Control (as applied to pest species)—reduction or maintenance of pest populations below the damage threshold.

Cornea—the outer transparent layer of the eye.

Cornicle—a tubular process (paired) which occurs on the abdomen of aphids.

Corpora allata—a pair of glands associated with the brain of insects which secrete juvenile hormone.

Coverage—the degree to which pesticide applied reaches the area intended. See ground cover, plant cover, target cover.

Coxa (pl. coxae)—the uppermost joint of insect legs by which they are attached to the thorax.

Crawler—the first instar nymph of scale insects which possesses legs and thus enables dispersal to take place.

Crochets—minute spines which form a pattern on the extremity of false legs of lepidopterous caterpillars.

Crop (as applied to insect anatomy)—the distended part of the foregut of insects that receives ingested food from the oesphagus.

Crop rotation—the practice of planting botanically unrelated crops on the same land in successive years.

Cross pollination—transfer of pollen between different plants of the same species.

Crustacea—a class of arthropods, mostly aquatic in habit, to which crabs and cray fish belong.

Crystalline cone—the clear conical shaped structure lying immediately beneath the cornea in each unit (ommatidium) of insect compound eyes.

Cultivar—(alternative term, variety)—cultivated variety of a plant.

Cultural control—manipulation of cultural practices to provide control of a pest.

Cuticle—the outermost non-cellular layer of an insect's body wall.

Damage threshold—pest population above which crop loss occurs.

Dead heart—the young caterpillar bores into the stem of the plant either from the top or side, eats it away and makes tunnel;s inside, as a result of which "dead hearts" appear.

Dengue—a virus disease of man marked by severe pain in head, eyes, muscles and joints and transmitted by certain mosquitoes.

Density dependent (factor)—a factor regulating populations of living organisms, whose influence varies with the population density of the organism concerned.

Density Independent (factor)—a factor regulating populations of living organisms, whose influence is independent of the population density of the organism concerned.

Deposit—the amount of an insecticide on a surface immediately following application.

Dermal toxicity—the toxicity of an insecticide (to higher animals) when applied to the intact skin.

Dermaptera—the order of insects comprising the earwigs.

Deterrent—a chemical substance which deters feeding or oviposition of an insect. cf. stimulant.

Detoxication—the process by which a poison is rendered harmless.

Deutonymph—the third instar stage in the development of mites and ticks. cf. protonymph.

Diapause—a state of suspended activity and metabolism which may occur at any stage in the life cycle of an insect as means of surviving unfavourable conditions. Diapause is normally brought about by the change in environmental conditions (especially reduced daylength) and requires a particular stimulus for its termination.

Dictyoptera—the order of insect comprising the cockroaches and mantids.

Diflubenzuron (Dimilin)—a chitin synthesis inhibitor; relatively non-toxic to mammals; especially effective against foliar feeding lepidopterous insects.

Dimorphism—a difference in bodily form, size or colour between two groups of individuals within a species.

Diploid—having the chromosomes present in homologous pairs, i.e., with a chromosome complement of 2n. cf. haploid.

Diplopoda—the order or arthropods within the class Myriapoda that comprises the millipedes.

Diplura—the order of insects that comprises the two-pronged bristle tails.

Diptera—the order of insects that comprises the true flies which are characterised by possession of only one pair of wings.

Dorsal—the top or upper (back) surface of an animal cf. ventral.

Drone—the male of social bees and wasps.

Dun—(alternative term, sub-imago)—winged but sexually immature mayfly adult which has to undergo another moult.

EC—see emulsifiable concentrate.

Ecdysis—(alternative term, moulting)— the process of shedding the skin (exoskeleton) undertaken periodically by insects and other arthropods.

Ecdysone—(alternative terms, moulting hormone, MH)—the hormone in insects which initiates the moulting process.

Economic control—control of a pest such that expenditure on control measures is more than compensated by increased value of yield.

Economic injury level—pest population level at which expenditure on control measure is just balanced by the value of increased yield resulting from control of the pest.

Economic threshold—(alternative term, action threshold)—pest population level at which control measures should be initiated to prevent the population rising above the economic injury level.

Ecosystem—the ecological system of a given area; composed of living plants and animals and non-living components.

Ectoparasite—a parasite which lives externally on the body of its host. cf. endoparasite.

Eelworm—see nematode.

Ejaculatory duct—the duct of male insects which conveys sperm to the copulatory organ.

Elytron (pl. elytra)—the hard horny forewings of beetles (Coleoptera) which act as protective cases for the membranous hind wings.

Embioptera—a small order of tropical insects which live gregariously in silken nests; sometimes referred to as web spinners.

Emulsifiable concentrate (EC)—an insecticide formulation in which the active ingredient is dissolved in a non-aqueous solvent to which emulsifiers are added. An emulsion is formed when added to water.

Encephalitis—inflammation of the brain; transmitted by *Aedes* sp.

Endemic (with reference to living organisms)—naturally occurring within a region.

Endemic typhus fever—a human disease caused by a bacterium like microorganism, *Rickettsia mooseri* and transmitted from rats to man by the oriental rat flea. Clinically similar to epidemic typhus fever except with milder symptoms.

Endoparasite—a parasite which lives internally within the body of its host. cf. ectoparasite.

Endopterygota—(alternative term, Holometabola)—the subdivision of insects that undergo complete metamorphosis with larva, pupa and adult stages. cf. Exopterygota.

Enzyme—a biochemical which promotes a specific chemical reaction.

Ephemeroptera—the order of insects comprising the mayflies.

Epidemic typhus fever—a human disease caused by *Rickettsia prowazekii* and transmitted by the body louse, *Pediculus humanus humanus.*

Eradication—Complete elimination of a species from an area or country.

Exarate (as applied to insect pupae)—having legs and other appendages separately encased (as in the pupae of beetles, bees, and wasps).

Exopterygota—(alternative term, Hemimetabola)—the subdivision of insects that undergo incomplete metamorphosis, cf. Endopterygota.

Exoskeleton—the external body wall of insects and other arthropods. Usually involves hardened plates over at least some of its area.

Family—the main unit of classification into which genera are grouped. Insect family names always end in-idae, e.g. Scarabaeidae.

Fat body—the principal food storage organ within the insect body; usually of diffuse and variable form providing "packing" around many internal organs.

Filariasis—a disease state caused by infection with filarial roundworms (nematodes) and transmitted by *Culex* sp. of mosquitoes.

Flabellum—the small spoon-shaped tip of the "tongue" of bees.

Float—a partial envelope, more or less inflated, around the anopheline eggs of mosquitoes.

Formulation—the form in which an insecticide is prepared for practical use.

Frass—insect excreta.

Fumigant—an insecticide which is sufficiently volatile to act at a distance in the vapour form.

Furca—a forked structure: as for example the forked skeletal support in the upper part of the mouthparts of flies.

Galea—the outer arm of the "pincers" present on the maxilla of insect mouthparts.

Gall—a swelling or outgrowth produced by a plant as a result of attacked by an insect, mite or micro-organism.

Ganglion (pl. ganglia)—a distinct swelling of nervous tissue in which many interconnections between nerve endings occur.

Gastric caeca—an outgrowth from the forepart of the mid-gut of insects.

Ganitalia—the external reproductive organs of insects (male or female).

Genus (pl. genera)—the next highest category to species in the classification of organisms.

Germ—the embryo within a seed.

Gizzard—the portion of the gut of insects immediately behind the crop; often thick-walled and muscular.

Glossa—the inner part of the extremity of the labium of insect mouthparts.

Gradual metamorphosis—alternative term for incomplete or partial metamorphosis.

Granaries—storage structures for the grain.

Granule—an insecticide formulation which consists of relatively large (1-3 mm) particles of an inert material (carrier) which are impregnated with insecticide or on which the insecticide is coated.

Gregarious—the habit of living together in groups.

Ground cover—the degree to which chemical applied covers the ground area being treated.

Grub—a scarabaeiform larva which is thickbodies with thoracic legs and well developed head; usually sluggish.

Grylloblattodea—a small order or primitive insects related to the cockroaches and crickets for which there is no common name.

Haemocoel—the body cavity of insects filled with blood (haemolymph) within which the internal organs lie.

Haemolymph—the unpigmented blood of insects.

Haltere(s)—the knob-shaped modified hind wings of Diptera which act as balancing organs.

Haploid—having a single set of unpaired chromosomes, i.e., with a chromosome complement of 1n. cf. diploid.

"Hard" pesticide—one which persists for a long time in the environment; particularly applies to organochlorine insecticides.

Head capsule—the hardened external skeleton of the head of insects.

Heart—the hindpart of the dorsal blood vessel of insects. The heart is muscular, equipped with inlet valves (ostia) and acts as a circulatory pump.

Hemelytron—the forewing of Hemiptera in which the basal portion is thickened and the distal portion membranous.

Hemimetabola—(alternative term, Exopterygota)—the subdivision of insects that undergo incomplete metamorphosis. cf. Holometabola.

Hemiptera—the order of insects that includes cicadas, leafhoppers, scale insects, aphids, plant bugs and many other groups; characterised by the possession of piercing and sucking mouthparts.

Herbivor—any animal (including insects) which feeds on living plants.

Hermaphrodite—having both male and female sexual organs in the same body.

Heteroptera—a sub-order of the insect order Hemiptera. The Heteroptera includes the true bugs. cf. Homoptera.

Hibernate—to enter a state of dormancy over the winter period. cf. aestivate.

High volume spraying—the application of pesticides in dilute form to thoroughly wet the object being sprayed; usually involves the application of 1120 or more liters per hectare (100 gals per acre).

Holometabola—(alternative term, Endopterygota)—the subdivision of insects that undergo complete metamorphosis. cf. Hemimetabola.

Homoptera—a sub-order of the insect order Hemiptera. The Homoptera includes cicadas, leafhoppers, aphids and scale insects. cf. Heteroptera.

Honeydew—excess sugary plant sap excreted from the hind end of the gut by sap sucking insects.

Hormone—a chemical, secreted into the bloodstream, which produces a specific effect on some bodily organ or process.

Host—a species of plant or animal on which an insect feeds.

Host plant—a species of plant on which a particular insect species feeds.

Hymenoptera—the order of insects that includes sawflies, ants, bees and wasps.

Hypermetamorphosis—an extreme form of metamorphosis undergone by some parasitic insects.

Hyperparasite—(alternative term, secondary parasite)—a parasite which attacks another insect which is itself a parasite.

Hypopharynx—a structure arising from the front surface of the labium in insect mouthparts and which acts as a tongue.

Hypopus—a special stage in the life cycle of some mites which is adapted for clinging to and being dispersed by other organisms.

IGR—see insect growth regulator.

Imago (pl. imagines)—the adult sexually mature form of an insect.

Immature stage—any stage of an insect prior to the imago (adult form).

Incompatible (as applied to pesticides)—not safe to mix together.

Incomplete metamorphosis—(alternative terms, partial metamorphosis, gradual metamorphosis)—metamorphosis in which there is a gradual, and often not very marked transformation from immature to adult stage. There is no pupal stage and immature stages are referred to as nymphs. cf. complete metamorphosis.

Indigenous—occurring naturally in country or area.

Inoculation—(alternative term, introduction)—(as applied to biological control). The introduction of new species of parasites, predators or disease organisms for biological control purposes.

Indigenous—occuring naturally in a country or area.

Inoculation—(alternative term, introduction)—(as applied to biological control). The introduction of new species of parasites, predators or disease organisms for biological control purposes.

Inhalation toxicity—the toxicity of an insecticide (to higher animals) when inhaled in the form of fine droplets or vapour.

Insect growth regulator (IGR)—group name applied to chemicals which interfere with normal insect growth or development, usually by affecting growth hormone systems.

Insectary—a place where insects are reared and bred.

Insect pathology—the study of insect diseases.

Insecticide—a chemical which is toxic to insects.

Instar—the form of an insect between moults.

Integrated control—(alternative terms, pest management, IPM)—the use of two or more different control measures together in an integrated fashion.

Integument—the outer body wall of an insect or other animal.

Introduction—see inoculation.

Invertebrate—any animal without a backbone (spinal column).

IPM—abbreviation of integrated pest management.

Isomer—a particular molecular arrangement of a chemical compound. Differrent isomers of the same substance have the same atoms within the molecule but their spatial arrangement differs.

Isoptera—the order of insects comprising the termites.

JH—an abbreviation of juvenile hormone.

Juvenile hormone—the hormone which retains an insect in its juvenile form; secreted by the corpora allata.

K-strategist—an ecological term applied to organisms which are adapted to ensure good survival. cf. r-strategist.

Key factor (as applied to population dynamics)—a factor which is dominant in regulating the numbers of an organism.

Key pest (s)—the serious and persistent pests which attack a crop.

Labium—the hind or lower lip of an insect's mouthparts.

Labrum—the front or upper lip of an insect's mouthparts.

Lac—resinous protective secretion of the lac insect.

Lacinia—the inner arm of the "pincers" present on the maxilla of biting insect mouthparts.

Larva—the active immature stage of insects which undergo complete metamorphosis. cf. nymph.

Larvicide—a chemical which kills the immature stages of an insect.

LD_{50}—dosage of a chemical which is lethal to 50 per cent of the test animals treated; usually expressed in mg per kg body weight.

Legal control (see quarantine)

Leishmaniasis—any disease caused by infection with a protozoan, *Leishmania* and transmitted by sand flies.

Lepidoptera—the order of insects comprising the butterflies and moths.

Life cycle (life history)—habits and changes undergone by an organism from the egg stage to its death as an adult.

Lindane—a synthetic chlorinated hydrocarbon containing 99 per cent or more of the gamma isomer of BHC.

Lint—the fiber surrounding the seed of unginned cotton.

Lorum—a basal sclerite present in the mouthparts of insects and well developed in bees.

Low volume spraying—the application of pesticides in concentrated form using about 225 to 560 litres per hectare (20-50 gals per acre).

Maggot—a vermiform larva without legs and without well-developed head capsule.

Malaria—an infectious disease of man caused by protozoa, *Plasmodium* and transmitted by mosquitoes of the genus *Anopheles*.

Male confusion technique—method of disrupting mating in insects by artificial release of excess amounts of a sex attractant chemical.

Mallophaga—the order of insects comprising the biting or bird lice.

Malpighian tubes—thread like organs opening into the beginning of the hind-gut of insects; excretory in function.

Mandible(s)—the second pair of appendages that make up the mouthparts of insects; in the form of biting jaws in biting/chewing insects.

Mandibulate—of biting/chewing pattern.

Mantle—the saddle-shaped swelling on the back of a slug.

Mass rearing—rearing of the immature stages in bulk.

Maxilla (pl. maxillae)—one of the third pair of appendages that make up the mouthparts of insects; usually complex in structure and bearing a sensory palp in biting/chewing insects but needle-like in piercing/sucking insects.

Mechanical control—any means of pest control which involves mechanically trapping or killing insects, or providing barriers to prevent insects gaining access to plants or other materials.

Mecoptera—a small order of insects comprising the scorpion flies.

Mentum—the central portion of the labium in insect mouthparts.

Mesothorax—the middle segment of the insect thorax.

Metabola—(alternative term, Pterygota)—the major subclass of insects which undergo metamorphosis whether complete or incomplete and which usually possess wings in the adult state. cf. Ametabola.

Metamorphosis—the change in bodily form which most insects undergo as they develop. See also complete metamorphosis, incomplete metamorphosis.

Metathorax—the hind segment of the insect thorax.

MH—(alternative terms, moulting hormone, ecdysone)—an abbreviation of moulting hormone.

Microbial control—the use of pathogenic micro-organisms to control pest species.

Micro-organism—a group term applied to living organisms that are in most stages too small to see with the naked eye; includes bacteria, viruses, fungi and protozoa.

Minimum tillage—techniques of sowing crops which involve little or no disturbance of the soil.

Mist blower—(alternative term, air blast sprayer)—a sprayer which uses a stream of air to carry spray droplets on to the object being sprayed.

Mite—a member of the class Acari related to but distinct from the insects. Mites may be distinguished from insects by the single body region, four pairs of legs and absence of antennae.

Miticide—(alternative term, acaricide)—chemical used for control of mites.

Mode of action (as applied to insecticides)—the mechanism by which an insecticide kills an insect.

Mollusca—the phylum of animals to which slugs and snails and their relatives belong.

Monocropping—(alternative term, monoculture)—the practice of growing the same crop (and often the same cultivar) on the same land year after year. In some situations it may involve the same crop plant over large areas.

Monoculture—see monocropping.

Monophagous—feeding on a single species of plant (or animal) or at least a group of closely related species. cf. polyphagous, oligophagous.

Morphology—the study of the bodily form of living organisms.

Moult, moulting—(alternative term, ecdysis)—the process of shedding the skin (exoskeleton) undergone periodically by insects and other arthropods.

Moulting hormone—(alternative terms, MH, ecdysone)—the hormone in insects which initiates the moulting process.

Mouth hooks—the paired hardened portions of the mouthparts of fly larvae which protrude from the mouth and enable them to shred their food.

Multivoltine—having several generations per year. cf. univoltine.

Myriapoda—a class of arthropods which includes millipedes and centipedes.

Nasute—a type of soldier caste in certain termites; this form bears a median frontal rostrum through which it ejects a defensive fluid; the jaws are small or vestigial.

Natural control—the natural regulation of populations of living organisms that takes place without human interference.

Natural enemy—any living organism which is harmful to a species by way of its predatory, parasitic or disease inducing habit.

Narrow spectrum (as applied to insecticides)—having an effect on only a narrow rang of insects.

Necrosis—death of a defined area of plant tissue.

Nectar—a sugary secretion produced by flowering plants which is highly attractive to insects.

Nectary—gland from which nectar is produced; normally located within the flowers but sometimes external to them.

Nematode—(alternative terms, eelworm, roundworm)—unsegmented smooth bodied worm belonging in the phylum Nematode; mostly minute in size.

Nematology—the study of nematodes.

Neuroptera—the order of insects that includes lacewings and ant lions.
Nit—the egg of a louse.
Nocturnal—active by night.
Non-preference—a mechanism of plant resistance to the pests whereby the pest chooses not to feed (or oviposit) on the resistant plant in comparison with susceptible ones.
Non-target organism—any organism other than that against which control measures (especially chemicals) are applied.
Nuptial flight—mating flight of a large swarm of males following one or more females prior to copulation; e.g., in bees and ants.
Nymph—the juvenile stage of an insect which undergoes incomplete metamorphosis. cf. larva.
Obligate parasite—a parasite which is obliged to follow a parasitic mode of life and which cannot exist in any other way.
Occasional pest—one which reaches significant levels only occasionally.
Ocellus (pl. ocelli)—a simple eye of insects and other arthropods incorporating a single lens.
Odonata—the order of insects comprising the dragonflies.
Oesophagus—the tubular forepart of the gut between the mouth and crop.
Oligophagous—feeding on a limited range of plants (or animals), usually within a single family. cf. monophagous, polyphagous.
Ommatidium (pl. ommatidia)—a unit of the insect compound eye.
Omnivorous—feeding on many different kinds of both plant and animal food.
Ootheca—an egg case or egg capsule of distinct form; as produced by cockroaches for example.
OP—abbreviation of organophosphate.
Opisthosoma—the hind body region of arachnids (spiders, mites, ticks).
Oral—via the mouth.
Oral toxicity— the toxicity of an insecticide (to higher animals) when ingested.
Order—a division of a class or sub-class composed of a group or groups of related families. The insect orders are the main insect types, e.g., order Coleoptera, beetles; order Lepidoptera, butterflies and moths.
Organochlorine (insecticide)—(alternative term, chlorinated hydrocarbon)—a group name applied to insecticides in which chlorine forms an important part of the molecule; DDT and lindane are examples.
Organophosphate (insecticide)—a group name applied to insecticides which have a phosphorus atom at the core of the molecule; malathion and parathion are examples.
Orthoptera—the order of insects that includes crickets, grasshoppers, locusts and wetas.
Osmeterium—a gland consisting of an eversible sac or pocket in which a repellent scent or other secretion collects and is used for defense by the moths.
Ostia (s. ostium)—valves by which blood enters the insect heart.
Ovariole—a unit or egg tube of the insect ovary.
Ovary— the female internal reproductive organ within which eggs are produced.
Oviduct—tube leading from the ovary and into which eggs are discharged.
Oviposition—the act of egg laying.
Ovipositor—specialised egg laying organ possessed by most female insects.
Palp—sensory feeler in the mouthparts of insects. There are usually two pairs of palps; one pair (maxillary palps) attached to the maxilla, and one pair (labial palps) attached to the labium.
Papilla—small, raised, finger-like process usually of minute size.
Paraglossa—the outer part of the extremity of the labium in insect mouthparts.
Parasite—an organism which lives in or on another.

Parasitoid—a term sometimes applied to insects which parasitise other insects (to distinguish them from parasites of vertebrate animals).

Parthenogenesis—reproduction without mating.

Partial metamorphosis—(alternative terms, incomplete metamorphosis, gradual metamorphosis)—metamorphosis in which there is a gradual and often not very marked transformation from immature to adult stage; there is no pupal stage and immature stages are referred to as nymphs.

Pathogen—an organism capable of causing disease in a plant or animal.

Penis—the male copulatory organ.

Peritrophic membrane—a membrane produced within the mid-gut of insects and which encloses the food as it passes along the alimentary canal.

Persistence—the degree to which the activity of an insecticide remains after it is applied.

Pest management—(alternative terms, integrated control, IPM)—an approach to pest control that blends biological and chemical control methods and involves application of insecticides according to needs, rather than according to a set programme.

Pest status—loss of 5 per cent or more in the yield of a crop.

Phasmida—the order of insects comprising stick and leaf insects.

Pheromone—a chemical substance produced by an insect or other animal which conveys a message to other members of the same species and usually results in behavioural response.

Phylum (pl. phyla)—major division of the animal kingdom corresponding to the main animal types, e.g., phylum Arthropoda—arthropods, phylum Mollusca—molluscs.

Physical control—pest control which involves either modification of some physical factor in the environment, (e.g., temperature) or use of a physical factor (e.g., light) to attract insects.

Physogastry—enlargement of the abdomen of female due to increased size of reproductive system, e.g., termites.

Phytophagous—plant feeding.

Phytotoxic—damaging to plant life.

Plant cover—the degree to which plants within a sprayed area receive a deposit of pesticide.

Plant host range—the species of plants on which an insect species feeds.

Plant resistance (to pests)—the ability of a plant to grow and produce economically despite the presence of a pest.

Plecoptera—the order of insects comprising the stoneflies.

Pollination—transfer of pollen from the male parts of flowers (anthers) to the female receptive organ (stigma).

Polyembryony—the production of several (often many) embryos from one egg.

Polymorphism—the existence of several distinctly different forms within the same species.

Polyphagous—feeding on a diverse range of plants (or animals). cf. monophagous, oligophagous.

Population dynamics—the study of changes in populations of organisms and of the reasons for such changes.

Potential pest—a pest which has the potential to be serious but which is usually suppressed by natural regulating factors.

Predator—an insect (or other animal) that captures and feeds on other animals (prey).

Pre-oviposition period—the period between mating and the commencement of egg laying.

Primary parasite (as applied to insects)—one which is parasitic on a plant feeding, predatory or scavenging species but never on another parasite.

Proboscis—(alternative term, rostrum)—the extended tubular mouthparts of an insect; particularly of the piercing/sucking pattern.

Propolis—a resinous material or bee-glue of vegetable origin used by bees in addition to wax for the building and repair of their nests.

Prosoma—the foremost body region of arachnids (spiders, mites, ticks) to which the walking legs are attached.

Prothoracic gland—a gland situated in the fore-part of the insect thorax which secretes moulting hormone.

Prothorax—the front segment of the insect thorax.

Protonymph—the second instar in the development of mites and ticks. cf. deutonymph.

Protura—a primitive order of small wingless insects for which there is no common name.

Pseudergate—last instar worker nymphs which retain the potential to moult into different castes.

Pseudopods—false legs on the abdomen of the larvae of butterflies, moths and sawflies.

Pseudo-tracheae—fine tubes which traverse the pad-like foot of the sponging mouthparts of flies.

Psocoptera—the order of insects comprising the book-lice.

Pterygota—(alternative term, Metabola)—the major subclass of insects which undergo metamorphosis whether complete or incomplete and which usually possess wings in the adult stage. cf. Apterygota.

Ptilinum—in Diptera an organ that can be inflated to a bladder-like structure and thrust out through frontal suture of the head at the time of emergence from the puparium.

Pupa (pl. pupae)—(alternative term, pupal stage)—the stage in the life cycle between larva and adult in insects which undergo complete metamorphosis.

Puparium (pl. puparia)—specialised form of pupa in which the last larval skin is retained as an extra covering; characteristic of the higher Diptera.

Pusa bin—a specialised storage structure for grains developed by scientists of Pusa Institute of New Delhi, in which a polythene sheet is inserted in the earthern cell structure.

Pyrethroid—class of insecticides with chemical structure similar to pyrethrum.

Quarantine—control of pests through legislation that enforces control measures or imposes regulations, to prevent the introduction or spread of pests.

Queen—female sexually reproductive form of social insects such as bees.

Queen substance—a secretion from the queen honey bee which inhibits sexual development of workers and regulates their building activities.

R-strategist—an ecological term applied to organisms which produce large numbers of offspring but with poor survival. cf. K-strategist.

Race—a naturally occurring group within a species usually visually indistinguishable but with some physiological difference from other members of the species, e.g., with a different plant host range, cf. strain.

Radula—the rasp-like tongue of slugs and snails.

Rectum—the hindmost portion of the gut of insects (and other animals).

Repellent—a chemical or physical source which induces insects to move away from it. cf. attractant.

Reproductive potential—(alternative term, biotic potential)—the maximum possible rate of increase of an organism in the absence of any limiting factors.

Residue—that portion of a pesticide which remains on or in produce at harvest.

Resistance (of insects to insecticides)—the developed ability of a strain within an insect species to withstand an insecticide to which it was formerly susceptible.

Resistance (of plants to insects)—see plant resistance.

Resurgence—the recovery of pest populations (sometimes to levels higher than before treatment) following application of an insecticide.

Retinula—the lower light-sensitive region of an ommatidium in the insect compound eye.

Rhabdom—the dense rod-like structure lying below the crystalline cone in each ommatidium of the insect compound eye.

Rickettsia—an obligate intracellular parasite of arthropods, many types of which are pathogenic for man and other animals. They are intermediate between bacteria and viruses.

Rocky Mountain spotted fever—a human disease caused by *Rickettsia rickettsi* and transmitted by ticks and bed bugs.

Rostrum—(alternative term, proboscis)—the extended forepart of the head of an insect cosisting of or bearing the mouthparts. The term is applied to insects such as weevils in which biting mouthparts are at the tip of a snout as well as to insects with piercing/sucking mouthparts.

Royal jelly—a secretion of pharyngeal glands of the worker honey bees.

Saprophagous—feeding on dead or decaying organic matter or on fungi.

Scavenger—an insect or other animal that feeds on dead and decaying plant and animal remains.

Sclerite—a hardened plate forming part of the exoskeleton of insects.

Secondary parasite—see hyperparasite.

Segment(s)—repeating body units of insects and of many other animals.

Selective insecticide—one which of effective against only a narrow range of insects.

Self pollination—transfer of pollen from anthers to stigma within the same flower.

Semilooper—a caterpillar with one or two pairs of the ventral prolegs missing; in crawling small loops of the body are formed.

Seminal vesicle—a storage reservoir for sperm in the reproductive system of male insects; usually paired.

Sericulture—rearing and breeding of the silk moth to obtain the raw silk fibre.

Sensillum—minute sensory structure on which insect's senses of smell and taste are based.

Seta (pl. setae)—slender body hair of filament.

Sex attractant—(related term, sex pheromone)—a volatile chemical substance produced by one sex of an insect (nearly always the female) to attract the opposite sex.

Sex pheromone—(related term, sex attractant)—a volatile chemical substance produced by one sex of an insect which produces some specific reaction in the opposite sex. Most sex pheromones are powerful attractants but male pheromones may act as aphrodisiacs.

Sex ratio—the ratio of males to females in a population.

Siphonaptera—(alternative term Aphaniptera)—the order of insects comprising the fleas.

Siphunculata—(alternative term, Anoplura)—the order of insects comprising the sucking lice.

Slug—a relative of snails but having the shell rudimentary or entirely wanting.

Snail—a member of the phylum Mollusca having a single, usually coiled shell and a broad, flat foot.

Social—living together in complex colonies; e.g., ants, bees.

Social hormone—a pheromone produced by the royal pair of termites and transferred throughout the colony by faecal trophollaxis. This presumably inhibits the production of secondary reproductives from pseudergates.

Soldier—a caste present in some social insects (especially termites and ants), whose task is to defend the colony.

Solitary—occurring singly or as mating pairs; not in colonies.

Soluble powder—an insecticide formulation which dissolves in water to form a solution.

Species—the lowest unit of classification normally used for plants and animals.

Spectrum of activity—the range of insect species against which an insecticide is active.

Spermatheca—a storage organ for sperm within the reproductive system of female insects.

Spermathecal gland—a gland associated with the spermatheca in the reproductive system of female insects.

Spermatophore—a package enclosing sperm which is transferred to the female during copulation.

Spinner—the fully mature adult form of mayflies after moulting of the dun (sub-imago) has taken place.

Spinneret—the central silk spinning bristle present in the mouthparts of caterpillars (larvae of Lepidoptera)

Spiracle—an external opening of the tracheal (respiratory) system of insects.

Stereomicroscope—a microscope with dual eye piece and objective lens systems which provides stereoscopic vision.

Sterile male technique—a method of pest control which involves the release of large numbers of sterile male insects.

Stigma—a dark mark near the fore-margin of the wings of many insects; a spiracle. Also (botanical), the female receptive part of a flower.

Stimulant—a chemical or physical factor which increases some specific insect activity (particularly feeding or oviposition). cf. deterrent.

Stipes—a basal sclerite forming part of the maxilla in the mouthparts of insects.

Stomach poison—an insecticide which has to be ingested by the insect before it exerts significant toxic action.

Strain—a group within a species that is in some way physiologically different from other members of the species, e.g., resistant to an insecticide. cf. race.

Strepsiptera—a small order of insects parasitic on other insects and commonly known as stylopids.

Stylet—a needle-like part of the mouthparts of insects and mites that feed with a piercing/sucking action.

Stylophore—the part of the head (capitulum) of mites that bears the stylets.

Sub-family—a taxonomic grouping below the level of family. Insect sub-family names always end in-inae, e.g., Melolonthinae.

Sub-imago—(alternative term, dun)—winged but sexually immature mayfly adult which has to undergo another moult.

Submentum—the basal portion of the labium in insect mouthparts.

Sub-oesophageal ganglion—the ganglion (swelling) of nervous tissue below the oesophagus in insects.

Sub-order—a taxonomic category below the level of order, e.g., sub-order Homoptera within the order Hemiptera.

Sub-species—a taxonomic category below the level of species.

Super-family—a taxonomic category immediately above that of family level.

Swath—the path treated by a single passage of a sprayer or other application equipment.

Symbiosis—association of two dissimilar organisms to their mutual advantage.

Symphilid—a member of the order Symphyla within the class Myriapoda.

Symphyla—an order of the class Myriapoda comprising the symphilids.

Symphypleona—a sub-order of springtails (Collembola) which are globular in form and without visible segmentation. cf. Arthropleona.

Synapse—branched connections between one nerve cell and another.

Synergist—a chemical which, when added to an insecticide, increases its effectiveness, but which is not itself toxic.

Systematics—(alternative term, taxonomy)—the study of the naming and classification of living organisms.

Systemic (as applied to insecticides)—having the property of being absorbed into and translocated by the sap stream of a plant or the bloodstream of an animal.

Target—object onto which a pesticide is to be deposited by spraying or other means.

Target cover—the degree to which the crucial portion (target) of a plant (or other sprayed object), receives a deposit of insecticide.
Tarsus (pl. tarsi)—the extremity of the insect leg and forming a "foot"; composed of one to five joints.
Taxonomy—(alternative term, systematics)—the study of the naming and classification of living organisms.
Tegmina (s. tegmen)—the leathery fore-wings of insects such as grasshoppers or cockroaches.
Teneral—newly moulted, pale and soft bodied.
Teratogenic—including birth deformities.
Termitaria—the nests build by termites under or above the ground.
Terrestrial—living on dry land.
Tertiary parasite (as applied to insects)—one which is parasitic on a secondary parasite.
Testis—the male internal reproductive organ which produces sperm.
Thorax—the middle of the three body regions of an insect.
Thysanoptera—the order of insects comprising the thrips.
Thysanura—the order of insects that includes three-pronged bristle tails such as silverfish.
Tick—a member of the order Acari within the class Arachnida; ectoparasitic on higher animals.
Tolerance (as applied to pesticides)—legally defined maximum permitted level of a pesticide on or in produce at harvest.
Tolerance (as applied to plant resistance to pests)—a mechanism of plant resistance to pests whereby the plant is able to grow and produce despite pest injury.
Trachea (pl. tracheae)—an air filled tube which forms part of the respiratory system of insects.
Transovarial—transmission via the egg from mother to offspring.
Tribe—a unit of classification below the family level into which groups of like genera are placed. Insect tribe names always end in-ini, e.g., Scarabaeini.
Trichoptera—the order of insects comprising the caddis flies.
Typhus fever—(see epidemic typhus fever).
Ultra low volume spraying (ULV)—the application of pesticides in extremely concentrated or undiluted form using 10 litres or less per hectare (1 gal per acre).
ULV—see ultra low volume.
Univoltine—having one generation a year. cf. multivoltine.
Varietal control—the use of varieties (cultivars) of cultivated plants which are resistant to pests.
Variety—a loosely defined taxonomic category below the level of sub-species. Usually possesses some visibly different character.
Vas deferens—the tube leading from the testis down which sperm passes.
Vector—an insect (or other animal) which transmits a disease causing micro-organism from one plant or animal to another, e.g., anopheline mosquitoes are vectors of malaria.
Vein (s)—the strengthening ribs in the wings of insects.
Ventral—the lower (front) surface of an animal. cf. dorsal.
Ventral tube— a tubular protrusion from beneath the first abdominal segment of springtails.
Vertebrate—higher animal possessing a backbone, e.g., mammals, birds, fish.
Vestigial—greatly reduced and non-functional.
Viviparity—the condition of giving birth to active young rather than laying eggs.
Waiting period—the prescribed period between last application of a pesticide to a crop and harvest.
Wettable powder (WP)—an insecticide formulation consisting of finely ground solid particles of active ingredient plus inert carrier; forms a suspension when added to water.
White earhead—the ears are dry and hollow, e.g., in wheat and rice.
Wind pollination—pollination in which pollen grains are carried by the wind rather than by insects.

Withholding period—the period of time following application of a pesticide, that livestock must be kept out of a pasture or fodder crop.

Worker(s)—sterile individuals in the nests of social insects which forage for food and generally run the colony.

WP—see wettable powder.

Yellow fever—an acute infectious disease caused by a virus transmitted by certain mosquitoes and marked by fever, jaundice, and albumin and globulin in the urine.

Yield forming organ—that part of a plant which gives rise to the harvested yield.

Zoraptera—a small order of tropical insects related to the book-lice for which there is no common name.